ELECTRICAL CONTRACTING
FORMS AND
PROCEDURES MANUAL

Other Books of Interest to Electrical Professionals

Anthony
ELECTRIC POWER SYSTEM PROTECTION AND COORDINATION

Cadick
ELECTRICAL SAFETY HANDBOOK

Croft & Summers
AMERICAN ELECTRICIANS' HANDBOOK

Hanselman
BRUSHLESS PERMANENT-MAGNET MOTOR DESIGN

Johnson
ELECTRICAL CONTRACTING BUSINESS HANDBOOK

Johnson
SUCCESSFUL BUSINESS OPERATIONS FOR ELECTRICAL CONTRACTORS

Kolstad
RAPID ELECTRICAL ESTIMATING AND PRICING

Kusko
EMERGENCY/STANDBY POWER SYSTEMS

Linden
HANDBOOK OF BATTERIES

Lundquist
ON-LINE ELECTRICAL TROUBLESHOOTING

Maybin
LOW VOLTAGE WIRING HANDBOOK

McPartland
HANDBOOK OF PRACTICAL ELECTRICAL DESIGN

McPartland
MCGRAW-HILL'S HANDBOOK OF ELECTRICAL CONSTRUCTION CALCULATIONS

McPartland
MCGRAW-HILL'S NATIONAL ELECTRICAL CODE HANDBOOK

Meland
ELECTRICAL PROJECT MANAGEMENT

Pete
ELECTRIC POWER SYSTEMS MANUAL

Richter & Schwan
PRACTICAL ELECTRICAL WIRING

Smeaton
SWITCHGEAR AND CONTROL HANDBOOK

Traister
SECURITY/FIRE ALARM SYSTEMS DESIGN, INSTALLATION, AND MAINTENANCE

ELECTRICAL CONTRACTING FORMS AND PROCEDURES MANUAL

RALPH EDGAR JOHNSON

GENE WHITSON

McGraw-Hill, Inc.

New York   San Francisco   Washington, D.C.   Auckland   Bogotá
Caracas   Lisbon   London   Madrid   Mexico City   Milan
Montreal   New Delhi   San Juan   Singapore
Sydney   Tokyo   Toronto

Library of Congress Cataloging-in-Publication Data

Johnson, Ralph Edgar.
    Electrical contracting forms and procedures manual / Ralph
Edgar Johnson, Gene Whitson.
        p.   cm.
    Includes index.
    ISBN 0-07-032699-1
    1. Electric contracting—Management.   2. Electric contracting—
Forms.   I. Whitson, Gene.   II. Title.
TK441.J643   1995
621.319´24—dc20   95-10809
                                                                CIP

1 2 3 4 5 6 7 8 9 0   KGP/KGP   9 0 0 9 8 7 6 5

ISBN 0-07-032699-1

*The sponsoring editor for this book was Harold B.
Crawford, the editing supervisor was Fred Dahl, and the
production supervisor was Pamela A. Pelton. It was set in
Caledonia by Inkwell Publishing Services.*

Printed and bound by Quebecor/Kingsport.

McGraw-Hill books are available at special quantity
discounts to use as premiums and sales promotions, or for
use in corporate training programs. For more information,
please write to the Director of Special Sales, McGraw-Hill,
Inc., 11 West 19th Street, New York, NY 10011. Or contact
your local bookstore.

This book is printed on acid-free paper.

Information contained in this work has been obtained by
McGraw-Hill, Inc., from sources believed to be reliable.
However, neither McGraw-Hill nor its authors guarantee the
accuracy or completeness of any information published
herein and neither McGraw-Hill nor its authors shall be
responsible for any errors, omissions, or damages arising out
of use of this information. This work is published with the
understanding that McGraw-Hill and its authors are
supplying information but are not attempting to render
engineering or other professional services. If such services
are required, the assistance of an appropriate professional
should be sought.

*This book is dedicated to the National Electrical Contractors
Association, a superb teammate of the electrical contracting
industry, whose untiring efforts have resulted in many years of a
progressive relationship and a strong, productive industry*

CONTENTS

Preface *ix*

CHAPTER ONE
Management *1*

CHAPTER TWO
Accounting *45*

CHAPTER THREE
Financial *85*

CHAPTER FOUR
Data Processing *95*

CHAPTER FIVE
Maintenance *105*

CHAPTER SIX
Estimating *115*

CHAPTER SEVEN
Job Management *179*

CHAPTER EIGHT
Job Preparation for Field *231*

CHAPTER NINE
Claims *253*

CHAPTER TEN
Purchasing *285*

CHAPTER ELEVEN
Warehouse *303*

Index *321*

PREFACE

Every business—large or small—benefits greatly from having good forms. For the electrical contractor who uses forms effectively, improvements are almost guaranteed in:

- Estimating.
- Allocating work units and man-hours.
- Bringing the right tools and materials to the work site on time.
- Identifying and accounting for unexpected costs brought on by clients' changes in specifications or schedule, or by unanticipated conditions on the job.
- Tracking payroll and other costs associated with jobs.
- Monitoring the overall financial condition of the business by means of a couple of basic financial reports.
- Controlling tool and material inventories.
- Overall coordination of activities in the field and in the office.

The myriad small details that characterize the electrical contracting business can be overwhelming. Preprinted forms, however, enable the contractor to organize each phase of the company's operations. With a good set of forms, no vital information is left out. The electrical contractor has all the information needed to submit an estimate, prepare a job, manage it efficiently, bring it to completion, and bill it promptly. Activities among the various departments and employees are better coordinated and more efficient, making more cost-effective use of the company working capital and man-hours. The contractor saves real dollars in the overall time spent on the completion of work tasks.

How This Book Is Organized

The *Electrical Contracting Forms and Procedures Manual* contains all the important forms used in electrical contracting. On the left-hand page for every form, you will find the following information:

- Form name
- Form number
- Originator
- Number of copies needed
- Recipients of the copies
- Suggested size
- Purpose

On the same page is an explanation of how the form is to be used. Where appropriate, the forms are partially or fully filled out, to show the information required. On the right-hand page is the form itself.

How to Use This Book

Many of the forms on the right-hand pages, such as organization charts, are filled in since they necessarily vary from one company to another; they are presented as examples only. Most forms, however, are blank, so that they can be copied or printed directly from the book and modified as necessary with your company's name, address, phone number, and logo—or used as is. Oversized forms (such as those requiring 8½×14-in. or 11×17-in. paper) can also be reproduced on a photocopier equipped with blow-up capability. Or your printer can blow them up prior to printing them. Blow-up percentages are noted in parentheses next to the SIZE entry on the left-hand page.

Alternatively, many forms, such as the Vehicle Accident Report or Sick Leave Request, can be entered into your word processor and modified with your company's data. This enables you to run off as many copies you need on any office laser printer.

Others can be created in a spreadsheet program, using the printed version as a model. For example, the Bid Summary Sheet can be related to the various takeoff schedules, so that it is automatically updated and retabulated as entries are made in the schedules.

No matter how you use the forms and procedures, they will help you in innumerable ways to run a more efficient and profitable contracting business.

Disclaimers

This book is designed to present accurate and authoritative information and to reflect the state of the law and regulations as they existed at the time of publication. Major changes may result from the future actions of code makers, labor legislators, Congress, state legislators, or the courts.

In recent years, the role of women in construction and engineering has seen tremendous growth, and this is indeed desirable. However, the terminology in these fields is replete with such words as *foreman, draftsman, workman,* and *craftsman.* In addition, the English language has no neuter personal pronoun. Therefore, where words such as *foreman* or the pronoun *he* are used, it is hoped that readers will understand them to refer to persons of either gender.

Ralph Edgar Johnson
Gene Whitson

CHAPTER ONE
MANAGEMENT

FORM NO.	FORM NAME	PAGE
M-1	Functional Organization Chart	2
M-2	Personnel Organization Chart	4
M-3	Sales Budget Calculation	6
M-4	Application of Overhead on Labor and Material	8
M-5	Application of Overhead by Department	10
M-6	Budget of Total Sales and Quotas for All Departments	12
M-7	Overhead Budget Summary	14
M-8	Memo	16
M-9	Request for Information	18
M-10	Safety Meeting Report	20
M-11	Accident Report	22
M-12	Estimating Man-Hour Register	24
M-13	Calculation of Period-to-Period Changes in Working Capital	26
M-14	Material Inventory	28
M-15	Material Requisition and Returned Material Record	30
M-16	Tool Inventory and Checklist	32
M-17	Tool Record Card	32
M-18	Estimating Man-Hour Register	36
M-19	Employee Wage Evaluation	38
M-20	Employee Personnel Record (Computer Screen)	40
M-21	Personnel Chart for the Marketing Department	42

NAME	# FUNCTIONAL ORGANIZATION CHART
NUMBER	**M-1**
ORIGINATOR	**General Manager**
COPIES	**At least one copy to each department**
DISTRIBUTION	**Manager of each department**
SIZE	**8½ × 11 in.**

PURPOSE

The phrase *functional organization* refers to a company that has shaped its management framework around the functions necessary to operate the business. In developing this management structure, the company groups related operations as parts of an important, defined business function. This system has the advantage of clearly defining the duties and responsibilities of each member of the management team, so that, when one or more of them leave the firm, the basic management framework remains intact and the specific duties of their successors are clearly defined.

An organization chart tailored to meet the specific requirements of a business is necessary to properly view the functional organization of that business. Form M-1 provides an illustration of such a chart.

FUNCTIONAL ORGANIZATION CHART

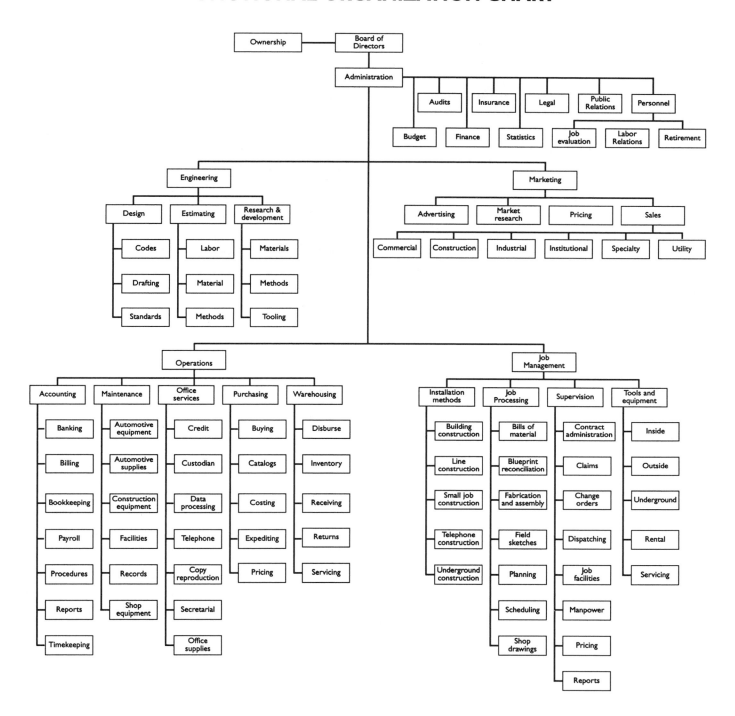

NAME	**PERSONNEL ORGANIZATION CHART**
NUMBER	**M-2**
ORIGINATOR	**Management**
NUMBER COPIES	**One to each department**
DISTRIBUTION	**All personnel**
SIZE	**8½ × 11 in.**

PURPOSE The personnel organization chart is developed from the functional organization chart. It shows people in their relationships to management and to one another: It defines the lines of authority and responsibility among employees. In a very small business, one person may perform many functions, and one or two persons may perform all the functions of the entire business. Regardless of the number of persons working in the business, all the basic functions must be performed at the time required.

PERSONNEL ORGANIZATION CHART

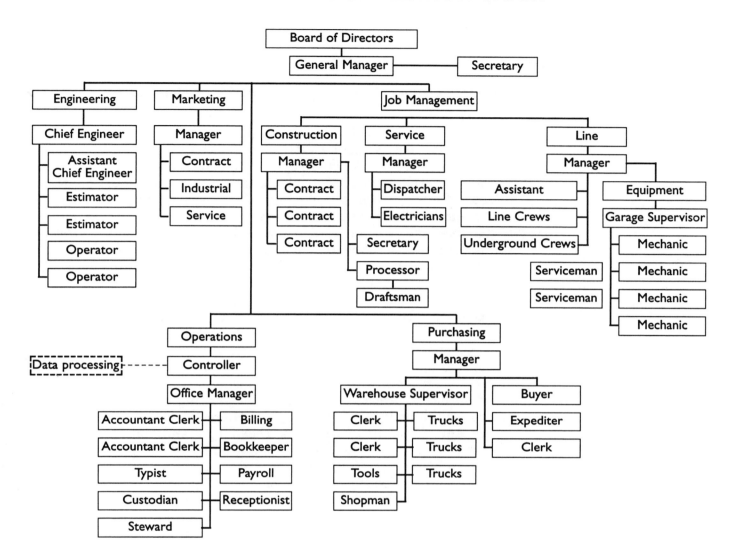

NAME	**SALES BUDGET CALCULATION**	
NUMBER	**M-3**	
ORIGINATOR	**Chief Accountant**	
COPIES	**4**	
DISTRIBUTION	**Accounting Office, General Manager, Assistant Manager, file**	
SIZE	**8½ × 11 in.**	

PURPOSE

Planning for a successful electrical contracting business necessarily starts long before the doors are open for business and the first contract is obtained. The amount of available financing capital must be determined. After preparing an opening balance sheet (which lists all the assets and liabilities of the business) is prepared, you should complete a reasonable budget of projected sales and anticipated profits, offset by estimated operating expense, along with required capital purchases.

The management of the new electrical contracting business will soon become aware that there is no such thing as true net profit on any one job due to overhead costs. A wide variety of jobs of different sizes and types is necessary to meet sales requirements. The most important responsibility of management is therefore to achieve the sales volume requirements of the business and to include in the billing for each job a proper proportion of the total overhead costs and anticipated profit. To track operations and achieve the business's established goals, the manager has to prepare a budget for the ensuing year's operations. In preparing the budget, the primary consideration is the determination of the amount of the current overhead cost and the sales volume required to cover the overhead and allow a reasonable profit margin.

Form M-3 illustrates a sales budget calculation for a small business projecting sales of $1 million for the first year of operation. Please note that the profit based on sales is only 2.06 percent and the estimated overhead is 19.38 percent of sales. The minimum working capital required is $100,000, and the net worth of the business should be at least $129,870.

SALES BUDGET CALCULATION

	Amount	Percent of Sales %
Material cost	365,800	36.58
Direct labor cost	273,700	27.37
Payroll tax (30% of labor)	82,110	8.21
Miscellaneous direct expense	32,300	3.23
Subcontract expense	31,400	3.14
Prime cost	785,310	78.53
Overhead	193,800	19.38
Profit	20,890	2.09
Sales	$1,000,000	100.00
Minimum working capital	100,000	*Sales*/10
Net worth (NW) 129,870		*Working Cap*/.77
Allowable building expense	14,285	11% of *NW*
Allowable truck expense	18,181	14% of *NW*
Allowable tool inventory	6,493	5% of *NW*
Allowable furniture and fixtures	12,987	10% of *NW*
Total fixed assets	51,946	40% of *NW*
Inventory allowance	12,987	10% of *NW*

Man-hours to be worked @ $14.00 per hour: 273,700/$14 = 19,500
Number of workers working an average of 50 weeks (2000 hours) = 19,500/2000 = 10

SALES BUDGET CALCULATION

	Amount	Percent of Sales %
Material cost		
Direct labor cost		
Payroll tax (30% of labor)		
Miscellaneous direct expense		
Subcontract expense		
Prime cost		
Overhead		
Profit		
Sales		
Minimum working capital		
Net worth (NW) 129,870		
Allowable building expense		
Allowable truck expense		
Allowable tool inventory		
Allowable furniture and fixtures		
Total fixed assets		
Inventory allowance		

Man-hours to be worked @   per hour:

Number of workers working an average of   weeks (hours) =

NAME	**APPLICATION OF OVERHEAD ON LABOR AND MATERIAL**
NUMBER	**M-4**
ORIGINATOR	**Chief Accountant**
COPIES	**4**
DISTRIBUTION	**Accounting Office, Estimation, General Manager, file**
SIZE	**8½ × 11 in.**
PURPOSE	Contract negotiations on different types of work may require contractors to quote a fixed labor rate, in which case the contractor quoting the lowest hourly rate will be awarded the contract. This situation most often arises when bidding on maintenance work for either a manufacturing facility, a commercial installation, or a hospital or medical complex.

In this situation, contractors should apply as much of their overhead as possible to material cost, which allows them to reduce their quoted hourly labor rate. This method of applying overhead results to create a lower labor rate is illustrated by Form M-4.

The budget is an educated projection of anticipated conditions during the period covered by the budget. If the conditions do not allow the company to meet the projections, it will be necessary to cut the overhead costs and then reevaluate the budget.

As the business expands into larger and more diverse types of contracts, such as line work, inside electrical work, and service work, it becomes necessary to develop different departments. This division into departments is required because of the different labor skills necessary to perform the tasks of the various types of work and the different materials installed in the various types of installation.

APPLICATION OF OVERHEAD ON LABOR AND MATERIAL

Labor Rate When All Overhead Is Applied on Labor

Labor Burden:

Overhead cost/man-hours = *193,800/19,500*	*$9.90* per hour	

Labor Rate:

Labor cost	*$14.00*
Payroll tax @ 30%	*4.20*
Prime cost	*$18.20*
Overhead	*9.90*
Profit *7.6%* of labor rate	*1.05*
Total	*$29.15* per man-hour

Labor Rate When 25% Is Added to Material Cost to Defray Part of Overhead

Budgeted material cost	*$365,800*	
Calculated 25% of material cost	*91,450*	
Budgeted overhead cost	*$193,800*	(*100%* of overhead)
Less: Amount added to material	*91,450*	(*47%* of overhead)
Overhead to add to labor	*$102,350*	(*53%* of overhead)

Labor Rate:

Labor cost	*$14.00*	
Payroll tax @ 30%	*4.20*	
Prime cost	*$18.20*	
Overhead	*5.25*	(*$102,350/19,500*)
Profit	*1.05*	
Total	*$24.50* per man-hour	

APPLICATION OF OVERHEAD ON LABOR AND MATERIAL

Labor Rate When All Overhead Is Applied on Labor

Labor Burden:

Overhead cost/man-hours =   per hour

Labor Rate:

Labor cost

Payroll tax @ 30%   _____

  Prime cost

Overhead

Profit   of labor rate   _____

  Total   _____ per man-hour

Labor Rate When 25% Is Added to Material Cost to Defray Part of Overhead

Budgeted material cost

Calculated 25% of material cost

Budgeted overhead cost   (% of overhead)

Less: Amount added to material   _____   (% of overhead)

Overhead to add to labor   _____   (% of overhead)

Labor Rate:

Labor cost

Payroll tax @ 30%   _____

  Prime cost

Overhead

Profit   _____

  Total   _____ per man-hour

NAME	**APPLICATION OF OVERHEAD BY DEPARTMENT**
NUMBER	**M-5**
ORIGINATOR	**Chief Accountant**
COPIES	**4**
DISTRIBUTION	**General Manager, Accounting Dept., Estimating Dept., file**
SIZE	**11 × 8½in.**

PURPOSE   When the departmental approach becomes necessary, management formulates the budget for each department in the same manner as for the single department, except that the budgeting for each department involves proration of overhead expenses according to its actual needs. For example, consider a company doing $5 million in sales and having three departments with entirely different labor burdens and overall makeup:

Department	Man Hours	Labor Burden
Department 1—Construction	170,400	?
Department 2—Service	37,275	?
Department 3—Line	149,100	?

Budgeting for these departments would be as shown on Form M-5. Notice the variances in the elements of cost among the departments. The Line Department furnished very little material for its work but rather installed material furnished by others. The Construction Department expended 48 percent of its total sales dollars for material compared to the 31 percent of total sales dollars expended for material by the Service Department.

The Construction Department and the Service Department each expended 32 percent of its sales dollars for labor, while the Line Department used just over 50 percent of sales for labor.

BUDGET FOR FISCAL YEAR 19____
APPLICATION OF OVERHEAD BY DEPARTMENT
SUMMARY OF ALL DEPARTMENTS

Cost	Department 1: Construction Budgeted Amount	%	Department 2: Service Budgeted Amount	%	Department 3: Line Budgeted Amount	%	Total Budgeted Amount	%
Material	$2,082,200	48.2	$269,700	31.0	$54,150	3.0	$2,406,050	34.4
Labor	1,382,400	32.0	278,400	32.0	911,520	50.5	2,572,320	36.7
Direct Job Expense	358,600	8.3	43,500	5.0	144,400	8.0	546,500	7.8
Equipment	13,200	0.3	25,800	3.2	347,540	19.2	386,540	5.5
Commissions	12,960	0.3	17,400	2.0	57,760	3.2	88,120	1.3
Total Direct Costs	$3,849,360	89.1	$634,800	73.2	$1,515,370	83.9	$5,999,530	85.7
Overhead	397,530	9.2	184,220	21.1	190,430	10.5	772,180	11.0
Profit	73,110	1.7	50,980	5.7	99,200	5.6	223,290	3.3
Total Sales	$4,320,000	100.0	$870,000	100.0	$1,805,000	100.0	$6,995,000	100.0
Man-hours	240,000		52,500		210,000		502,500	
Labor burden	$1.66 per man-hour		$3.51 per man-hour		$0.91 per man-hour			

BUDGET FOR FISCAL YEAR 19____
APPLICATION OF OVERHEAD BY DEPARTMENT
SUMMARY OF ALL DEPARTMENTS

Cost	Department 1: Construction		Department 2: Service		Department 3: Line		Total	
	Budgeted Amount	%	Budgeted Amount	%	Budgeted Amount	%	Budgeted Amount	%
Material								
Labor								
Direct Job Expense								
Equipment								
Commissions								
Total Direct Costs								
Overhead								
Profit								
Total Sales								
Man-hours								
Labor burden	per man-hour		per man-hour		per man-hour			

NUMBER	**M-6**
ORIGINATOR	**Sales Manager**
COPIES	**4**
DISTRIBUTION	**Sales Manager, General Manager, Assistant Manager, file**
SIZE	**8½ × 11 in.**

PURPOSE Once the budget summary for all departments is prepared, a method should be devised to outline the production or sales required of each department to attain the budgeted goal. Form M-6 illustrates the establishment of sales quotas for each salesperson by department. As the budget period progresses, this form becomes the "yardstick" by which each salesperson and consequently each department is progressing toward the goals established by the budget.

When preparing the budget of total sales and quotas for each salesperson, the application of overhead to each type of work should be determined as nearly as possible to actual costs per man-hour. For example, the overhead costs for the Service Departments will be much higher than the Construction or Line Department due to:

✓ The larger number of jobs.

✓ The necessity of buying material in smaller quantities

✓ The need for more vehicles.

BUDGET OF TOTAL SALES AND QUOTAS FOR ALL DEPARTMENTS

Department 1			Man-Hours	Sales Budgeted
Salesperson's quotas:				
Bud Quinn			80,000	$1,440,000
Dick Thompson			15,000	270,000
John Hallinan			22,000	396,000
Dwight Johnson			93,000	$1,674,000
			210,000	$3,780,000
Plus: Change Orders			30,000	540,000
Total man-hours			240,000 @1.66	
Total sales				$4,320,000

Department 2	Man-Hours Residential	Commercial	Industrial	Sales Budgeted
Salesperson's quotas:				
Bob Lindgren	6,000	3,000	3,500	$207,200
John Hallinan		8,000		132,560
Doc Easley		1,000	10,000	182,270
Dick Thompson	1,500	10,000	1,000	207,125
	7,500	22,000	14,500	$729,155
Plus: Service department	2,000	4,000	2,500	140,845
Total man-hours	9,500 @3.80	26,000 @2.15	17,000 @1.85	
Total sales				$870,000

Department 3			Man-hours	Sales Budgeted
Salesperson's quotas:				
Dick Thompson			50,000	430,000
Doc Easley			120,000	1,032,000
Dwight Johnson			40,000	343,000
Total hours and sales			210,000 @2.85	$1,805,000
Total sales all departments				$6,995,000

BUDGET OF TOTAL SALES AND QUOTAS FOR ALL DEPARTMENTS

Department 1	Man-Hours	Sales Budgeted
Salesperson's quotas:		
	———————	———————
Plus: Change Orders	———————	———————
Total man-hours		
Total sales		

	Man-Hours			
Department 2	Residential	Commercial	Industrial	Sales Budgeted
Salesperson's quotas:				
	———————	———	———————	———————
Plus: Service department	———————	———	———————	———————
Total man-hours				
Total sales				

Department 3	Man-hours	Sales Budgeted
Salesperson's quotas:		
	———————	———————
Total hours and sales		
Total sales all departments		

NAME **OVERHEAD BUDGET SUMMARY**

NUMBER **M-7**

ORIGINATOR **Chief Accountant**

DISTRIBUTION **Accounting Office, General Manager, Sales Department, file**

SIZE **11 × 8½ in.**

PURPOSE To ensure the desired amount of profit, a tight control must be exercised not only over the gross amount of sales but, equally important, also over spending on overhead expense items. At the end of each month, a trial balance is generated by the computer, which will list the expenditures to date on each item of overhead expense. These totals may then be transferred by computer to the Overhead Budget Summary for comparison with budgeted amounts.

✓ Column 1 on the form lists the overhead item by account number and name.

✓ Column 2 indicates the total amount charged to the item to the date of the report.

✓ Column 3 lists the total annual budgeted amount for each item.

✓ Column 4 indicates the percentage amount of the budgeted amount allowed to date from the beginning of the accounting period to the date of the report. For example, the first month of the period would be indicated as 8.3 percent, the second month as 16.6 percent, and so on.

✓ Column 4 lists the dollar amount of the budgeted amount allowed to date.

✓ Column 6 is a calculation of the total expended to date, subtracted from the allowed amount of budgeted expense shown in column 5. A negative amount in this column indicates that actual expenditures are exceeding the budget.

✓ Column 7 indicates the percentage amount over or under budget for each item.

OVERHEAD BUDGET SUMMARY
MONTH OF JANUARY 19XX

(1) A/C # Name	(2) Total to Date	(3) Total Budget	(4) Budget % to Date	(5) Budgeted Amount to Date	(6) Dollar Amount (Over) or Under Budget	(7) % (Over) or Under Budget
501 Advertising	$720	$8,735	8.33	$728	$8	1.05
502 Auto and truck	$1,900	$22,463	8.33	$1,871	($29)	-1.54
503 Bad debts	$980	$9,723	8.33	$810	($110)	-13.59
504 Charitable contributions	$150	$1,796	8.33	$150	($0)	-0.26
505 Collection expense	$20	$106	8.33	$9	($11)	-186.51
506 Depreciation and amortization	$2,000	$25,236	8.33	$2,102	$102	4.86
507 Dues and subscriptions	$250	$3,226	8.33	$269	$19	6.97
508 Education expenses	$90	$1,101	8.33	$92	$2	1.87
509 Overhead employee benefits	$1,500	$13,405	8.33	$1,117	($383)	-34.33
510 Freight and express	$42	$559	8.33	$47	$5	9.30
511 Heat, light, power, and water	$476	$5,168	8.33	$4,310	($46)	-10.57
512 Company insurance—general	$1,765	$19,511	8.33	$1,625	($140)	-8.60
514 Overhead employee workers compensation	$910	$10,927	8.33	$910	$0	0.08
515 Public and employers liability insurance	$275	$4,324	8.33	$360	$85	23.65
516 Legal and accounting	$1,565	$17,529	8.33	$1,460	($105)	-7.18
518 Miscellaneous	$296	$4,426	8.33	$369	$73	19.71
520 Office supplies and expense	$1,065	$12,716	8.33	$1,065	$0	0.01
521 Pension and profit-sharing plans	$1,500	$20,398	8.33	$1,699	$199	11.78
522 Rent	$2,050	$24,612	8.33	$2,050	$0	0.01
523 Repairs and maintenance	$604	$7,277	8.33	$606	$2	0.36
525 Plan, bid bonds, estimating, and engineering	$295	$3,353	8.33	$279	($16)	-5.62
526 Salaries and wages—overhead personnel	$22,910	$275,021	8.33	$22,910	($0)	.00
528 Shop supplies and expense	$367	$3,153	8.33	$321	($46)	-14.35
530 Small tools	$456	$5,470	8.33	$456	$0	0.11
532 Taxes and licenses—general	$187	$10,213	8.33	$157	($30)	-3.55
534 Taxes—overhead payroll	$1,731	$20,161	8.33	$1,731	($0)	-0.08
536 Telephone, telegraph, and postage	$910	$11,434	8.33	$952	($21)	-3.29
537 Travel—administrative	$536	$6,435	8.33	$536	$0	0.01
550 Other expenses	$612	$7,313	8.33	$614	$2	0.31
Totals	$46,949	$558,348	8.33	$46,502	($447)	-0.96

All the dollar amounts are totaled and the percentages of the totals are shown at the bottom of the form. By preparing this report on a monthly basis, management has an early warning of any potential trouble areas in overhead expenditures and is able to make necessary adjustments. Form M-7 has been completed as a sample for a contractor performing approximately $5 million in sales per year.

OVERHEAD BUDGET SUMMARY
MONTH OF JANUARY 19XX

(1) A/C # Name	(2) Total to Date	(3) Total Budget	(4) Budget % to Date	(5) Budgeted Amount to Date	(6) Dollar Amount (Over) or Under Budget	(7) % (Over) or Under Budget
Advertising						
Auto and truck						
Bad debts						
Charitable contributions						
Collection expense						
Depreciation and amortization						
Dues and subscriptions						
Education expenses						
Overhead employee benefits						
Freight and express						
Heat, light, power, and water						
Company insurance—general						
Overhead employee workers compensation						
Public and employers liability insurance						
Legal and accounting						
Miscellaneous						
Office supplies and expense						
Pension and profit-sharing plans						
Rent						
Repairs and maintenance						
Plan, bid bonds, estimating, and engineering						
Salaries and wages—overhead personnel						
Shop supplies and expense						
Small tools						
Taxes and licenses—general						
Taxes—overhead payroll						
Telephone, telegraph, and postage						
Travel—administrative						
Other expenses						
Totals						

NAME	**MEMO**
NUMBER	**M-8**
ORIGINATOR	**Varies**
COPIES	**One to all necessary parties**
DISTRIBUTION	**As required**
SIZE	**8½ × 11 in.**
PURPOSE	The best communications are those conducted by means of documentation. This eliminates the misunderstandings that often occur in verbal communication and also establishes the what, where, when, and how of a situation. It is always difficult for one person to clearly relate detailed information to another. In the construction industry, the passing of information from one person to another presents a particularly complicated and often disconcerting problem.

The best communications are those conducted by means of documentation. This eliminates the misunderstandings that often occur in verbal communication and also establishes the what, where, when, and how of a situation. It is always difficult for one person to clearly relate detailed information to another. In the construction industry, the passing of information from one person to another presents a particularly complicated and often disconcerting problem.

In the electrical contracting business, the problem becomes more difficult because many individuals, from the engineering designer to the field installer, must perform functions in a logical sequence and then pass information on to other people. The physical separation of personnel further complicates the process. Due to these factors, communication must be complete, clear, concise, consistent, and simple. Sample form M-8 is one simple style of intercompany memo that can be used.

MEMO

DATE:_____

TO: _____

FROM: _____

SUBJECT: _____

Job Number:_____

Job Name:_____

Drawing Number: _____

Action Required: YES _____   NO _____

When Action Is Required: Date _____ / _____ / _____   Time _____

Required Action:

Message:

_____

Memo Received [date]: _____ / _____ / _____

Signed: _____

NAME	**REQUEST FOR INFORMATION**
NUMBER	**M-9**
ORIGINATOR	**Engineering Department**
COPIES	**5**
DISTRIBUTION	**Architect, Engineer, General Foreman, file**
SIZE	**8½ × 11 in.**

PURPOSE

Design standards are the common language used throughout the organization. If correctly used and conscientiously applied, they will allow consistent communications that interrelate all the required functions and operations.

Since design is fundamental to the entire system and defines the basic methods to be employed in making the detailed installation, the most up-to-date drawings must be provided to everyone involved immediately upon execution. Each of the drawings represents a typical construction assembly and provides the means of communication throughout the construction process.

The primary design of the job determines the pattern to be followed in the field installation. The type of material required for the installation is determined by the type of construction. Research and development in the electrical industry has resulted in many materials that are competitive in price, readily available, easy to install, and yet as good as materials used in the past.

Please note that Form M-9 refers to the Job Name, Reference Drawing Number (Ref. Drg. #), the Grid Lines on the reference drawing (Ref. Grid Lines #), and the Specifications Section Number (Spec. Section #). The form also notes the firm to which it is addressed, the name of the individual who should handle the request, the date the information is required, and the name of the sender. The routing for distribution is shown, along with space for the date the response is issued and the signature of the person replying to the request.

REQUEST FOR INFORMATION

License Number: _____

JOB NAME: _____   DATE: ____ / ____ / ____   Request for Information Number: _____

TO: [Firm] _____   REF. DRG.#:_____

ATTENTION: _____   REF. GRID LINES #: _____

SPEC. SECTION #: _____

Request:

_____

_____

_____

_____

_____

_____

_____

_____

_____

_____

_____

_____

_____

_____

Date Response Needed: ____ / ____ / ____

Copies to: _____   By: _____

_____

Reply:

_____

_____

_____

_____

_____

_____

_____

_____

_____

_____

_____

_____

_____

NAME	**SAFETY MEETING REPORT**
NUMBER	**M-10**
ORIGINATOR	**Safety Officer**
COPIES	**5**
DISTRIBUTION	**Safety Officer, General Manager, General Foreman, bulletin board, file**
SIZE	**8½ × 11 in.**

PURPOSE | On-the-job injuries may be the most costly item in the construction industry. An injured worker is covered under the worker's compensation laws of all states, and an employer's workers compensation insurance premiums are based on the number and the seriousness of injuries. In most cases, the insurance premiums are based on the calculation covering the past three years, so that an injury today affects the employer's costs for three successive years.

It is therefore imperative that employers do everything possible to lower the accident rates on their jobs. To accomplish this, they must install regular formal safety training programs and require all employees to attend them.

The most advantageous place to hold safety training sessions is on the job site. These meetings should be held weekly and should be conducted by the general foreman as a direct representative of top management.

Form M-10 provides management with a report of the safety meeting and lists the date held, the job where it was held, the safety points covered at the meeting, and the names of the employees who attended the meeting. The completed forms should be kept as a permanent record in the files of the company Safety Officer.

SAFETY MEETING REPORT

DATE:____ / ____ / ____   JOB: _____

MEETING CONDUCTED BY: _____

Safety Points Covered

_____

_____

_____

_____

_____

_____

Attendants' Names

NAME	**ACCIDENT REPORT**
NUMBER	**M-11**
ORIGINATOR	**General Foreman**
COPIES	**5**
DISTRIBUTION	**Safety Director, treatment center, personnel file, job file, Foreman**
SIZE	**5½ × 8½ in. (Copy form at 100% of original; cut it out along box rules.)**

PURPOSE

Form M-10, Safety Meeting Report, and Form M-11, Accident Report, have been included in the management form section because of the importance of controlling the accident rate and subsequent costs in any business. The accident report should be completed for every injury, whether it was serious or not. The person or persons responsible for insuring safe practices for all employees in the company must examine the circumstance surrounding any serious injury to employees. This examination enables them to take the steps necessary to avoid that type of injury in the future.

The report covering any serious accident contains the pertinent information that must be reported to the carrier of the company's workers compensation insurance. The accident report and a copy of the insurance company report should be filed in the employee's personnel record for referral in case future claims are filed by the employee.

The Safety Director should compile a report describing all the accidents that took place during the current period, noting the names of the job sites where they occurred and preventative measures to be taken to avoid other accidents. This report should be presented at the next supervisors meeting and discussed at length.

ACCIDENT REPORT

DATE OF REPORT:____ / ____ / ____

EMPLOYEE NAME: _____

DATE ACCIDENT OCCURRED:____ / ____ / ____

Employee's Occupation at Time of Accident: _____

_____

Where Did the Accident Happen? _____

_____

Description of the Injury: _____

_____

_____

Treatment Facility Referred to: _____

_____

What Could Have Been Done to Avoid This Accident? _____

_____

_____

_____

_____

_____

_____

_____

Will This Be a Lost Time Accident? Yes _____   No _____

Names of Witnesses to Accident:

_____

_____

_____

_____

Report Prepared by: _____

Signed by Foreman: _____

NAME	**ESTIMATING MAN-HOUR REGISTER**
NUMBER	**M-12**
ORIGINATOR	**Chief Estimator**
COPIES	**4**
DISTRIBUTION	**Chief Estimator, Sales Department, General Manager, Estimating personnel file**
SIZE	**8½ × 11 in.**

PURPOSE

The continuing health of the business requires good estimates in sufficient volume to ensure the receipt of enough sales to produce the budgeted profit. To achieve this, quotas must be established for the Estimating Department.

Electrical contractors who make certain that they secure a volume of work that generates enough dollars to cover costs and a reasonable profit will be successful. To do this it is necessary to:

✓ Prepare realistic budget projections of the volume of work required for the full year in terms of man-hours.

✓ Allocate the required budgeted man-hours as quotas to the estimator or estimating team.

✓ Maintain a management control register that shows the number of man-hours budgeted versus the man-hours of work awarded. The same register should show the total man-hours estimated and comparison with a quota of work to be estimated. The types of jobs being bid will affect the comparison, but the register will illustrate to the manager how much work is being processed and by whom.

The suggested register is shown as Form M-12. The use of this form enables the manager to determine the percentage of quota estimated and the percentage of quota awarded. (See also M-18, page 36.)

ESTIMATING MAN-HOUR REGISTER

WEEK NO: _____

DATE: _____

MAN-HOURS OF WORK: _____

Department	Name	Man-hours of work estimated						Man-hours of work awarded				
		Quota This Week	Estimated This Week	Previous Estimate	Total to Date	Quota to Date	% Quota	Quota This Week	Quota to Date	Awarded This Week	Awarded to Date	% Quota
	Totals											

NAME	**CALCULATION OF PERIOD-TO-PERIOD CHANGES IN WORKING CAPITAL**
NUMBER	**M-13**
ORIGINATOR	**Chief Accountant**
COPIES	**3**
DISTRIBUTION	**General Manager, Chief Accountant, file**
SIZE	**8½ × 11 in.**

PURPOSE   To ensure the ongoing growth and success of the business, management must control the financial affairs of the business by:

1. Controlling sales and working capital by:
   a. Ascertaining the sales volume available to the company in the market available to it and in the type of work it is familiar with.
   b. Setting up a proper amount of working capital to support that volume of sales.
   c. Setting an obtainable goal of a volume of sales based on the company's margins of overhead and profit.
2. Controlling the elements of working capital as follows:
   a. *Net worth:* Having sufficient capital invested in the business.
   b. *Profit:* Making a profit in the operation of the business.
   c. *Fixed Assets:* Allocating the investment of capital in proper proportions as follows: tool inventory allowance, 5 percent of net worth; material inventory, 10 percent of net worth; furniture and fixtures, 10 percent of net worth; building rent, 11 percent of net worth; truck rental, 14 percent of net worth.
   d. *Long-term liabilities:* Considering borrowing capital for at least a five-year payout to shore up working capital and expand sales.

Operating the business within the limits imposed by the financial condition of the company is absolutely necessary to assure the successful continuation and growth of the business. This requires the perpetual examination of accounting reports and the willingness to make corrections in operations promptly when required to maintain the proper balance of the financial elements. Form M-13 is an aid in the calculation of the changes in working capital. (Other financial control forms are illustrated in the chapter on financial forms.)

**CALCULATION OF PERIOD-TO-PERIOD
CHANGES IN WORKING CAPITAL**

Source of Working Capital from Operations:		
Net earnings		$55,650
Depreciation (no cash outlay)		9,800
Working capital from operations		$65,450
Proceeds from long-term borrowing		$55,450
Cash value of life insurance		$31,100
Sale of land held for investment		7,600
Sale of fixed assets		3,800
		$100,950
Gross increase in Working Capital		$166,400
Decreases in Working Capital:		
Additions to equipment	$39,950	
Payment on long-term debt	12,100	
Increased cash value of life insurance	3,050	
		$55,100
Current increase in working capital		$111,300
Working capital at beginning of period		99,300
Total working capital at end of period		$210,600
Increases in Current Assets:		
Cash and certificates of deposit		$53,100
Accounts receivable		100,300
Underbillings		29,000
Other current assets		13,100
		$195,500
Increases in Current Liabilities:		
Accounts payable	$30,000	
Overbillings	13,450	
Income taxes	38,300	
Other current liabilities	2,450	
		$84,200
Working Capital (Current Assets − Current Liabilities)		$111,300

CALCULATION OF PERIOD-TO-PERIOD CHANGES IN WORKING CAPITAL

Source of Working Capital
  from Operations:

    Net earnings

    Depreciation (no cash outlay)

    Working capital from operations

    Proceeds from long-term borrowing

    Cash value of life insurance

    Sale of land held for investment

    Sale of fixed assets

Gross Increase in Working Capital

Decreases in Working Capital:

    Additions to equipment

    Payment on long-term debt

    Increased cash value of life insurance

Current increase in working capital

Working capital at beginning of period

    Total working capital at end of period

Increases in Current Assets:

    Cash and certificates of deposit

    Accounts receivable

    Underbillings

    Other current assets

Increases in Current Liabilities:

    Accounts payable

    Overbillings

    Income taxes

    Other current liabilities

Working Capital (Current Assets − Current Liabilities)

NAME	**MATERIAL INVENTORY**
NUMBER	**M-14**
ORIGINATOR	**Accounting Department**
COPIES	**2**
DISTRIBUTION	**Accounting, Purchasing**
SIZE	**8½ × 11 in.**

PURPOSE
The Material Inventory is considered a current financial asset, and it is therefore necessary to take a physical inventory or a count of all material at least annually. This listing is then priced out at cost, and the Accounting Department adjusts the financial records of the company to reflect the actual dollar amount of the inventory.

Maintaining continuity and accuracy requires a standard method of listing and pricing the items of material in the inventory.

Upon completion of the count and pricing procedure, a copy of the form is forwarded to the Purchasing Department to be used to identify items of material that are in inventory and to avoid duplication when buying new material. Any items of material in the inventory that are determined to be surplus and that are in original packages should be returned to the vendor, whenever possible, for credit.

The annual (or more often, if necessary) inventory record is the basic information necessary for control of the material inventory.

MATERIAL INVENTORY

DATE: _____ / _____ / _____

Quantity	Description	Catalog Number	Unit Price	Cost

Total _____

NAME	**MATERIAL REQUISITION AND RETURNED MATERIAL RECORD**
NUMBER	**M-15**
ORIGINATOR	**Warehouse Department**
COPIES	**3**
DISTRIBUTION	**Cost Accounting, warehouse, job site**
SIZE	**8½ × 11 in.**

PURPOSE The value of controlling inventory lies in the fact that funds tied up in material inventory cannot be applied to essential uses of working capital. Material inventory is considered a current asset, but, as with the fixed assets, the values involved reduce working capital by an equivalent amount.

Therefore the value of the material inventory must be controlled to an amount equal to and not exceeding 10 percent of net worth. If a business has accurate inventory records, it can finance a large percentage of the inventory through a bank on a short-term loan, which, if handled properly, may increase the functions of working capital as a long-term liability.

With careful planning and the use of computers, it is now possible to predict optimum inventory levels and to establish procedures and controls to maintain such levels. Excess stock must be returned to vendors for credit whenever possible.

Form M-15, Material Requisition and Returned Material Record, provides a means for the warehouse to make proper entries to the inventory control records.

While the tool inventory is not considered a current asset on the financial statements, the amounts invested in tools as a fixed asset have a direct bearing on the amount of working capital available to finance the work projects. The value of the tool inventory may be equal to but not exceed 5 percent of net worth.

A perpetual tool inventory should be maintained, and a physical inventory of tools should be taken at least once a year. The annual physical inventory could be prepared on a form such as Form M-16, and the perpetual inventory should be maintained on a card such as Form M-17. The Tool Record Card is also used to record all the maintenance performed on the tool. When an established percentage of maintenance is required to keep the tool in operating order, the tool can be replaced.

TOOL INVENTORY AND CHECKLIST

DATE:____ / ____ / ____   JOB: _____   Page __ / __

Tool	On Hand	Tool	On Hand

TOOL RECORD CARD

Description	Size Model or Type	Mfg. Serial Number	Shop Serial Number	Date Purchased	Purchase Price

Purchased from		Address		Telephone	

Condition						

Date	Repairs or maintenance	Parts Cost	Labor Cost	Total Cost

NAME	**ESTIMATING MAN-HOUR REGISTER**
NUMBER	**M-18**
ORIGINATOR	**Accounting Department**
COPIES	**4**
DISTRIBUTION	**General Manager, Chief Accountant, Chief Estimator, file**
SIZE	**8½ × 11 in.**

PURPOSE

The accountant refers to overhead costs as "General and Administrative Expense," but in the electrical construction industry these costs are generally known as "overhead." Overhead costs are fixed costs, and they accrue whether or not any sales are made. The costs are controllable within certain limits, but once management has committed to a purchase or lease of real estate, for example, these costs are seldom subject to change.

The simplest way to apply overhead costs is to apply a fixed percentage to the prime cost of each job, provided that the prime cost of every job includes the same percentage of material, direct labor, and direct job expense. It is unlikely that the exact relationship of prime costs will occur very often, but the one thing that will occur on every job is man-hours of labor.

The control of overhead expense and its allocation as a billable job expense is one of the most important controls in the electrical contracting business. For control and allocation of overhead, management needs the following:

✓ Accurate budget of overhead applicable to each department.

✓ Accurate forecast of man-hours to be worked by each department.

✓ Adequate record to indicate the degree of accomplishment of work projected. The Estimating Man-Hour Register, Form M-18, may be used as the basis for man-hour information. (See also Form M-12, page 24.)

ESTIMATING MAN-HOUR REGISTER

DATE: _____

WEEK NO: _____

MAN-HOURS OF WORK: _____

Department	Name	Man-hours of work estimated						Man-hours of work awarded				
		Quota This Week	Estimated This Week	Previous Estimate	Total to Date	Quota to Date	% Quota	Quota This Week	Quota to Date	Awarded This Week	Awarded to Date	% Quota
	Totals											

NAME	**EMPLOYEE WAGE EVALUATION**
NUMBER	**M-19**
ORIGINATOR	**Middle Management**
COPIES	**4**
DISTRIBUTION	**Personnel File, General Manager, middle management, Payroll**
SIZE	**11 x 8½ in.**

PURPOSE

Managers must realize that people are the company's greatest asset. They must help the people who work in the organization to pursue their own interests, as well as to fulfill their needs and aspiration. This interest in their personal well-being must be expressed not only through tangible means like wages, benefit plans, and the availability of programs for self-development and graining, but also by creating the cordial, constructive, and comfortable working relationships that will encourage people to function as a team.

To many employees, although the actual wage may be only part of the compensation package, it is the most important because it determines the individual's standard of living. Therefore, management must protect the employees' interests by establishing a wage policy that is fair to all concerned, in balance with the economy, and nondetrimental to employee morale. To do this it is necessary to establish a base wage for each position and to set standards to be used in evaluating employee performance.

The *base wage* for each position in the company should be determined by a study of the community standard wage for comparable positions and, if possible, the base wage your competitors are paying for this position. Most of the wage studies conducted by the states or local communities will indicate low, median, and high wages for each job category. Remember that these studies include input from a broad spectrum of businesses and therefore probably do not provide an accurate standard for the electrical contracting business. Because of this, the wages noted in the various studies should be used only as a guideline in establishing your base wage rates.

There are five basic elements to be considered when evaluating an employee's job performance (as shown in Form M-19). Such a form greatly simplifies the evaluation process for the manager or supervisor. The completed form should become a permanent part of the employee's personnel file and be referred to each time an evaluation is performed.

The evaluations should be undertaken for each employee at regularly scheduled times at least one a year. It is extremely important that the results of the evaluation be discussed with the employee and that the employee be advised of the reasons for the points assigned to each item, as well as what the employee can do to improve the score in future evaluations.

EMPLOYEE WAGE EVALUATION

EMPLOYEE NAME: _____ JOB TITLE: _____ DATE: ___ / ___ / ___

Factor	Descriptive Point Levels					Reasons	Points
	20	15	10	5	1		
Quality	Ability to work to plans and/or specifications						
	Excellent Always works to spec. & under difficult conditions	**Good** Rarely does poor work	**Fair** Occasionally does poor work	**Poor** Frequently does poor work	**Very Poor** Very often does poor work		
Reliability	Compliance with company policies, procedures, attendance, promptness, conventional appearance, personal hygiene, safe work habits						
	Excellent Always in compliance with all requirements	**Good** Rarely fails to comply	**Fair** Normal compliance	**Unreliable** Often undependable	**Very Unreliable** Completely undependable		
Quantity	Produces to standard labor unit requirement						
	Excellent Consistently exceeds standard	**Good** Frequently exceeds standard	**Fair** Occasionally meets standard	**Poor** Seldom produces at standard rate	**Very Poor** Consistently turns out very little work		
Attitude	Willingness to comply, cooperate, get along with supervisers and fellow workers						
	Excellent Always complies with objectives, gets along with other workers	**Good** Usually complies, cooperates, gets along with others	**Fair** Occasionally complies and has trouble with others	**Poor** Frequently reacts begrudgingly, does not get along well	**Very Poor** Generally uncooperative and difficult to work with		
Job knowledge	Knowledge, experience, continuing education, skills level						
	Excellent Performs perfectly with little assistance	**Good** Performs most aspects of job and rarely requires assistance	**Fair** Quite often requires assistance on complex operations	**Low** Performs limited part of job or requires close supervision	**Very low** Fairly unskilled and requires close supervision		

TOTAL POINTS _____

Evaluated by _____ Job title _____ Department _____

Date for next review ___ / ___ / ___

NAME	**EMPLOYEE PERSONNEL RECORD (COMPUTER SCREEN)**
NUMBER	**M-20**
ORIGINATOR	**PERSONNEL OFFICE**
COPIES	**1 copy (Record is maintained in computer.)**
DISTRIBUTION	**Employee's personnel file**
SIZE	**8½ × 11 in.**
PURPOSE	The Employee Personnel Record is primarily used to establish information for payroll purposes. The initial record to be set up for payroll use will contain the following information:

1. Employee's name
2. Employee's company payroll number
3. Employee's address and Social Security number
4. Employee's pay rate (either hourly or periodic) and applicable tax information
5. Employee's status (hourly or salary)
6. Employee's job classification
7. Number of dependents and other personal information
8. Earned vacation classification
9. Sick leave classification
10. Other benefit information
11. Union benefit deductions and other deductions
12. Employee's date of hire
13. Any other items required

The Employee Personnel Record is maintained in the computer, and changes to any of the information may be made as necessary to keep the record current.

EMPLOYEE PERSONNEL RECORD

1. EMPLOYEE NO.: _____   2. NAME: _____

                                                    3. ADDRESS: _____

                                                    4. CITY/ST/ZIP: _____

5. SSN:____ ____ ____   6. TELEPHONE: ( )-____-_____

Sex	EEOC	Dept.	Marital Stat.	W/H Dep.	St.Tax	Exempt.	Fixed Fed.	Driver

16. Pay Code: _____   17. Hourly Rate:_____   18. Salary:_____

19. Union Code: _____   20. Profit Share Y/N: _____   21. Birth Date:____ ____ ____

22. Hire date:_____   23. Term Date: _____

24. Car Allowance: _____

25. Vacation Code: _____   26. Vacation Hours Accrued:_____   27. Sick Hours Accrued: _____

Deductions

Code	Description	Amount	Limit
_____	_____	_____	_____
_____	_____	_____	_____
_____	_____	_____	_____
_____	_____	_____	_____
_____	_____	_____	_____

	Gross	Federal	OASDI	Medicare	State
MTD	_____	_____	_____	_____	_____
QTD	_____	_____	_____	_____	_____
YTD	_____	_____	_____	_____	_____

	Regular	OT	DT	Other	
Hours	_____	_____	_____	_____	_____ Net
Amount	_____	_____	_____	_____	_____ Gross
					_____ Federal

Check Date: ____ / ____ / ____   _____ OASDI

Check Number: _____   _____ Medicare

Check Type: _____   _____ State

Sick Hours: _____   _____ Ded. 1

Vacation Hours: _____   _____ Ded. 2

                                                                 _____ Ded. 3

                                                                 _____ Ded. 4

                                                                 _____ Ded. 5

Display	Next	Previous	Enter	Change	Remove	Help	Quit

NAME	**PERSONNEL CHART FOR THE MARKETING DEPARTMENT**
NUMBER	**M-21**
ORIGINATOR	**Sales Manager**
COPIES	**3**
DISTRIBUTION	**General Manager, Sales Manager, file**
SIZE	**8½ × 11 in.**

PURPOSE It is always the responsibility of management to determine the dollar amount of sales necessary to produce enough income for the company to cover all expenses and allow for a reasonable profit. To accomplish this the manager must develop a comprehensive budget and establish sales quotas that will enable the company to reach its goals.

Salespersons must have incentives to function properly. Setting a required volume of sales provides one incentive. Properly compensating the person who achieves it provides another. Both are necessary.

Sales quotas are also fundamental to budgeting; after they have been established, it is a clerical operation to compare monthly the quota with the actual accomplishment. This comparison must be made in an organized manner by salespersons and by type of sales, as illustrated as Form M-21.

PERSONNEL CHART FOR MARKETING DEPARTMENT

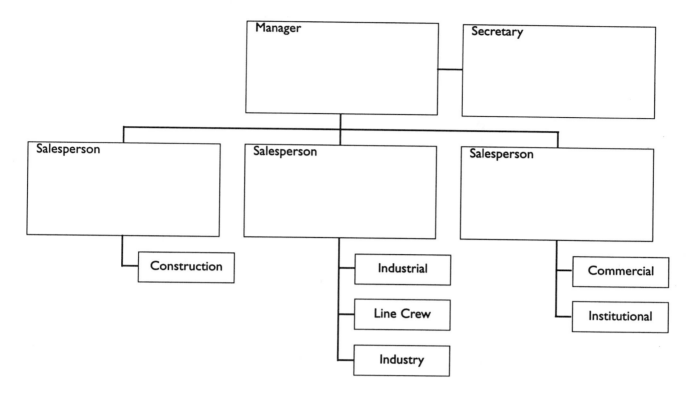

CHAPTER TWO
ACCOUNTING

FORM NO.	FORM NAME	PAGE
A-1	Accounts Payable Vendor Master (Computer Screen)	46
A-2	Accounts Payable Invoice Register (Computer Screen)	48
A-3	Accounts Payable Invoice Register	50
A-4	Check Register	52
A-5	Payables by Job	56
A-6	General Ledger Report	58
A-7	Job Cost Master (Computer Screen)	60
A-8	Invoice	62
A-9	Accounts Receivable Job Master (Computer Screen)	64
A-10	Sales Journal	66
A-11	Time and Material Billing Register	68
A-12	Accounts Receivable Aging Report	70
A-13	Cash Receipts Journal	72
A-14	Schedule of Fixed Assets	74
A-15	General Ledger (Computer Screen)	76
A-16	General Ledger Chart of Accounts	78
A-17	Weekly Payroll (Computer Screen)	80
A-18	Payroll Register	82

NAME	**ACCOUNTS PAYABLE VENDOR MASTER (COMPUTER SCREEN)**
NUMBER	**A-1**
ORIGINATOR	**Accounting Department**
COPIES	**2**
DISTRIBUTION	**Accounting, Cost Accounting, a permanent file in the computer**
SIZE	**8½ × 11 in. on the computer screen**

PURPOSE An Accounts Payable Vendor Master containing the information shown on Form A-1 should be set up in the computer for each vendor. During the posting of information to the Accounts Payable input screen, the entry of the vendor ID number automatically creates a vendor's name entry on the Accounts Payable Invoice Register (see Form A-2).

An Accounts Payable Vendor Master should be completed each time a company purchase order is issued to a new vendor. This should be a required standard procedure without exception because it establishes the base for an audit trail of expenditures.

ACCOUNTS PAYABLE VENDOR MASTER

1. Vendor ID No.: _____

2. Vendor Name: _____

3. Remit Address: _____

_____

6. Contact Name: _____

_____

7. Telephone: (_____) _____-_____

8. Date Estb.: ____ ____ ____

9. Last Active: ____ ____ ____

10. Our Account Number: _____

11. YTD Purchase: _____

NAME	**ACCOUNTS PAYABLE INVOICE REGISTER (COMPUTER SCREEN)**
NUMBER	**A-2**
ORIGINATOR	**Accounts Payable Department**
COPIES	**1**
DISTRIBUTION	**Accounts Payable, permanent file in the computer**
SIZE	**8½ × 5½ in.**

PURPOSE   The Accounts Payable Invoice Register records every invoice or other billing received by the company. The information entered on the computer entry screen will be used to:

✓ Complete the Accounts Payable Invoice Register.

✓ Record the invoice amounts to the proper General Ledger accounts (see Form A-15).

✓ Set up the allowable cash discount.

✓ Pay the accumulated amounts due each vendor when the checks are drawn.

ACCOUNTS PAYABLE INVOICE REGISTER

Vendor Number: _____   Vendor Invoice Number: _____   Invoice Date:____ / ____ / ____

Invoice Due Date:____ / ____ / ____   Invoice Terms: _____   Discount, %:____

Job Number: _____   Job System Number: _____   Invoice Amount: $_____.____

General Ledger Account Number:_____.____

Display   **Next**   **Previous**   **Enter**   **Change**   **Remove**   **Help**   **Quit**

NAME **ACCOUNTS PAYABLE INVOICE REGISTER**

NUMBER **A-3**

ORIGINATOR **Accounts Payable Department**

COPIES **I**

DISTRIBUTION **Accounts Payable**

SIZE **14 × 11 in. (Copy form at 125% of original.)**

PURPOSE The Accounts Payable Invoice Register should contain the following information:

1. Vendor or supplier name and ID or code number
2. Vendor or supplier invoice number
3. Invoice date
4. Invoice due date
5. Job number
6. Job system number
7. General Ledger account number
8. Invoice amount
9. Discount amounts available

All this information is used for valuable reports during and after the current accounting period.

It is advantageous to list the vendor by a *code number* rather than by the full name because of the keystrokes saved in inputting the information.

When the vendor's code number is input, the computer is enabled to write the check to the proper vendor when the invoice is processed for payment, apply the invoice amount to job cost by job number, and post the invoice amount to the correct general ledger account.

The vendor or supplier's *invoice number* is required to establish an audit trail of receipts of material or services and payments made on all invoices entered into the system. This number will be entered by the computer on the voucher of each check written to the supplier and used to verify payment.

The *invoice date* and *invoice due date* will be used to further identify the invoice and establish the company's payment date. The due date of the invoice will be used to determine the company's required cash flow for any given period.

The *job number* entry allows the computer to distribute the cost indicated on the invoice to the applicable job. The *job system number* allows a further sort to apply the cost to a particular system or division of each job as established on the estimate. All this information then becomes part of job cost for cost accounting purposes.

ACCOUNTS PAYABLE INVOICE REGISTER

Vendor Number	Vendor Name	Invoice Number	Invoice Date	Due Date	Invoice Amount	Discount Amount	Job Number	Job System Number	GL Account Number
1001	ABC Supply	2465	03/10/XX	04/10/XX	$45,623.00	$456.23	P31090	P31090-3	120.1
1125	B&J Supply	69801	03/07/XX	04/10/XX	545.30	10.90	P20791	P20791-1	120.1
1746	Greg Supply	14207	03/26/XX	04/25/XX	1,234.54	24.69	P21091		120.1
1803	Haps Mechanical	1004	03/03/XX	04/10/XX	225.45	2.25	STOCK		121.0
2130	J&J Electric Supp	2965	03/17/XX	04/10/XX	2,540.91	50.81	P30191	P30191-6	120.1
2645	New Supply Co.	24356	03/04/XX	04/10/XX	1,567.32	15.67	P21091		120.1
2750	Owl Office Equip.	18902	03/08/XX	04/10/XX	104.98	1.05			525.0
2750	Owl Office Equip.	18945	03/12/XX	04/10/XX	35.00	0.35			525.0
2845	Perfect Electric	32654	03/02/XX	04/10/XX	5,675.00	$113.56	P20791	P20791-1	120.1
4545	Walts Electric	1587	03/12/XX	04/10/XX	254.95		STOCK		121.0

ACCOUNTS PAYABLE INVOICE REGISTER

Vendor Number	Vendor Name	Invoice Number	Invoice Date	Due Date	Invoice Amount	Discount Amount	Job Number	Job System Number	GL Account Number

NAME	**CHECK REGISTER**
NUMBER	**A-4**
ORIGINATOR	**Accounts Payable Department**
COPIES	**2**
DISTRIBUTION	**Accounts Payable, Cost Accounting**
SIZE	**14 × 11 in. (Copy form at 125% of original.)**

PURPOSE The Check Register is a list of checks drawn by the computer to apply to the liability set up by the accounts payable amount on the Invoice Register. The check register contains the following information:

1. Company check number
2. Vendor or supplier name and ID number
3. Vendor's or supplier's invoice number
4. Gross amount of invoice
5. Discounts allowed
6. Discounts taken
7. Net amount of check

The information shown on the check register is sorted by vendor and compiled by the computer from the invoice information input on the Accounts Payable Invoice Register screen.

The *company check number* entry serves three purposes: (1) controlling the checks issued by the company, (2) verifying payment of invoices, and (3) checking cleared checks against the company's bank statements for reconciliation purposes.

The *vendor's* or *supplier's name* and *invoice number* identify the payment and allow vendors and suppliers to properly apply the payment on their records.

The *gross,* or total, *amount* of the invoice is the amount that will be debited against the accounts payable account in the General Ledger and used for balancing the various vendor or supplier accounts payable in the subsidiary accounts payable ledger, which lists all entries made to each vendor's account.

The *discounts allowed* entry establishes the total amount of discount available if all accounts are paid according to the invoice terms.

The *discounts taken* entry indicates the amount of discount that has been deducted from the total amount of the invoice being paid.

The computer deducts the discount taken from the gross amount of the invoice and draws a check for the *net amount.* The discount taken amount is applied to an income account in the General Ledger, and the net amount of the check is deducted from the bank account in the General Ledger. This amount is used when reconciling the bank account to bank statements.

At the end of a period, the computer, by sorting all payments made to a vendor and then adding the current total to the previous period as totals, determines the total amount of business the vendor has received from the company for the year-to-date period. This information will be posted to the vendor master automatically each time invoices are entered. In some instances this information may be used in negotiations for better service or prices. At other times it can be used to determine a more even distribution of purchase orders to a group of suppliers.

Note that Form A-4 illustrates the company's checks listed on the register in numerical order, which facilitates reconciliation of the company bank account with the bank's monthly statement. In addition, the computer sorts all due accounts payable invoices in alphabetical order by vendor. If more than one invoice is due to a vendor, only one check is drawn to cover the total for all the invoices.

CHECK REGISTER

Check Number	Vendor Code Number	Vendor Name	Invoice Number	Invoice Amount	Discount Allowed	Discount Taken	Check Amount
6000	1001	ABC Supply	1001	$45,623.00	$456.23	$456.23	$45,166.77
6001	1125	B&J Supply	69801	$ 545.30	$ 10.90	$ 10.90	$ 534.40
6002	1746	Greg Supply	14207	$ 1,234.54	$ 24.69	$ 24.69	$ 1,209.85
6003	1803	Haps Mechanical	1004	$ 225.46	$ 2.25	$ 2.25	$ 223.21
6004	2130	J&J Electrical Sup	2968	$ 2,540.91	$ 50.81	$ 50.81	$ 2,490.10
6005	2645	Now Electric Suppl	24356	$ 1,567.32	$ 15.67	$ 15.67	$ 1,551.65
6006	2780	Owl Office Equip	18902				$ 0.00
			18945	$ 139.98	$ 1.40	$ 1.40	$ 138.58
6007	32654	Perfect Electric	32654	$ 5,678.00	$113.56	$113.56	$ 5,564.44
6008	1587	Walts Electric	1587	$ 254.95	$ 0.00	$ 0.00	$ 254.95
							$ 0.00
Total				$57,809.46	$675.51	$675.51	$57,133.95

CHECK REGISTER

Check Number	Vendor Code Number	Vendor Name	Invoice Number	Invoice Amount	Discount Allowed	Discount Taken	Check Amount
Total							

	NAME	**PAYABLES BY JOB**

NAME **PAYABLES BY JOB**

NUMBER **A-5**

ORIGINATOR **Accounts Payable Department**

COPIES **2**

DISTRIBUTION **Cost Accounting, Accounts Payable**

SIZE **8½ × 11 in.**

PURPOSE All the invoices received by the company from vendors and suppliers are listed on the Invoice Register not only to set up a payment schedule but also to enable the computer to sort the invoices by job number and General Ledger account number. This procedure permits the printing of a listing of accounts payable invoices by company job number, as illustrated on Form A-5.

The Payables by Job report is used by the Cost Accounting Department as an audit list to verify the sources of charges made to each job for the accounting period.

PAYABLES BY JOB

Job Number	Expense Amount	Vendor	Vendor Name	Invoice Number	Invoice Date	Liability Date	GL Account	
1393	495.72	2100	Brown Wholesale	419946	02/11/XX	03/31/XX	120	1
Job Total	495.72							
1493	14.00	1100	Accurate Engraving	3324	02/26/XX	03/31/XX	120	1
	95.00	3600	Concrete Coring Company	188954	03/10/XX	03/31/XX	120	1
	93.00	16345	Phoenix Electric Supply Co.	364849001	03/04/XX	03/31/XX	120	6
	447.81	16345	Phoenix Electric Supply Co.	364852501	03/11/XX	03/31/XX	120	1
	1,265.20	2100	Brown Wholesale	424824	03/05/XX	03/31/XX	120	1
	25.64	2100	Brown Wholesale	428740	03/16/XX	03/31/XX	120	1
	950.00	3600	Concrete Coring Company	188994	03/17/XX	03/31/XX	120	6
	255.00	16345	Phoenix Electric Supply Co.	364849008	03/04/XX	03/31/XX	120	1
	81.98	23650	Woertz	3863	03/10/XX	03/31/XX	120	1
Job Total	3,227.03							
21293	644.00	2100	Brown Wholesale	419953	02/12/XX	03/31/XX	120	1
	159.04	19350	Simplex Time Recorder Co.	32354058	02/24/XX	03/31/XX	120	1
Job Total	803.04							
21392	377.00	2100	Brown Wholesale	428436	03/22/XX	03/31/XX	120	1
Job Total	377.00							
21393	11,356.00	5450	Electro Test Inc.	46002in.	02/26/XX	03/31/XX	120	6
Job Total	11,356.00							
21493	2,430.30	2100	Brown Wholesale	419952	02/12/XX	03/31/XX	120	1
Job Total	2,430.30							
21693	.00	2100	Brown Wholesale	419945	02/11/XX	03/31/XX	120	1
	292.50	2100	Brown Wholesale	420867	02/10/XX	03/31/XX	120	1
	40.00	2100	Brown Wholesale	424145	03/02/XX	03/31/XX	120	1
Job Total	332.50							
21793	23.05	12280	Littlepage Office Supplies	86165	03/03/XX	03/31/XX	120	5
	-95.91	2100	Brown Wholesale	1544	03/29/XX	03/31/XX	120	1
	495.60	2100	Brown Wholesale	423925	03/23/XX	03/31/XX	120	1
	1,003.50	2100	Brown Wholesale	423924	02/11/XX	03/31/XX	120	1
	1,011.81	2100	Brown Wholesale	423927	03/09/XX	03/31/XX	120	1
	412.00	2100	Brown Wholesale	423929	02/16/XX	03/31/XX	120	1
Job Total	2,850.05							
21893	171.00	269	City Treasurer	3024866	03/19/XX	03/31/XX	120	5
	1,750.00	999	Arizona Pacific Wood Preserv.	1304	03/09/XX	03/31/XX	120	5
	-770.00	999	Bowens Poles	1305	03/03/XX	03/31/XX	120	1
	770.00	999	Bowens Poles	1309	03/03/XX	03/31/XX	120	1
	604.46	2100	Brown Wholesale	428997	02/25/XX	03/31/XX	120	1
	40.00	1550	Arizona Wholesale Supply Co.	0119704	03/24/XX	03/31/XX	120	1
	115.90	7310	G.E. Supply	699624	03/04/XX	03/31/XX	120	1
	150.23	19925	Sunstate Equipment Corporation	236791001	03/18/XX	03/31/XX	120	5
	26.95	20250	24th Street Lumber Co.	08915	03/01/XX	03/31/XX	120	5
	244.82	2100	Brown Wholesale	425926	03/11/XX	03/31/XX	120	1
	1,386.00	2100	Brown Wholesale	426892	03/12/XX	03/31/XX	120	1
	4,340.00	2100	Brown Wholesale	426893	03/08/XX	03/31/XX	120	1

PAYABLES BY JOB

Job Number	Expense Amount	Vendor	Vendor Name	Invoice Number	Invoice Date	Liability Date	GL Account

Job Total

Job Total

Job Total

Job Total

Job Total

Job Total

Job Total

Job Total

	NAME	**GENERAL LEDGER REPORT**
	NUMBER	**A-6**
	ORIGINATOR	**Accounts Payable Department**
	COPIES	**1**
	DISTRIBUTION	**Accounting Department**
	SIZE	**8½ × 11 in.**

PURPOSE Form A-6, General Ledger Report, is produced by the computer from information entered on the Invoice Register and lists all the current period invoices by applicable General Ledger account number.
This listing is used as an audit list to verify the sources of charges to the various General Ledger expense and inventory accounts.

GENERAL LEDGER REPORT

Account Number	Expense Amount	Vendor	Vendor Name	Invoice Number	Invoice Date	Liability Date
119 0	−50.00	920	Varney, Sexton, Lunsford, Aye	1301	03/01/XX	03/31/XX
	50.00	920	Varney, Sexton, Lunsford, Aye	1302	03/25/XX	03/31/XX
	−25.00	999	Deveny Architects	1302	03/01/XX	03/31/XX
	−100.00	999	Sullivan Durano, Inc.	1306	03/01/XX	03/31/XX
	−50.00	999	Devenny Architects	1310	03/01/XX	03/31/XX
	50.00	999	St. Joseph Hospital	1316	03/25/XX	03/31/XX
Acct Total	−125.00					
120 1	19.37	439	Hinckley & Schmidt Co. of Az.	4272855	03/30/XX	03/31/XX
	−770.00	999	Bowens Poles	1305	03/03/XX	03/31/XX
	770.00	999	Bowens Poles	1309	03/03/XX	03/31/XX
	14.00	110	Accurate Engraving	3324	02/26/XX	03/31/XX
	495.72	2100	Brown Wholesale	411147	12/28/XX	03/31/XX
	15.96	2100	Brown Wholesale	416694	01/27/XX	03/31/XX
	602.08	2100	Brown Wholesale	417910	01/29/XX	03/31/XX
	51.62	2100	Brown Wholesale	419927	02/08/XX	03/31/XX
	.00	2100	Brown Wholesale	419945	02/11/XX	03/31/XX
	495.72	2100	Brown Wholesale	419946	02/11/XX	03/31/XX
	2,430.30	2100	Brown Wholesale	419952	02/12/XX	03/31/XX
	644.00	2100	Brown Wholesale	419953	02/12/XX	03/31/XX
	292.50	2100	Brown Wholesale	420267	02/10/XX	03/31/XX
	10.17	2100	Brown Wholesale	421899	02/19/XX	03/31/XX
	69.80	2100	Brown Wholesale	421916	02/23/XX	03/31/XX
	134.99	2100	Brown Wholesale	421921	02/23/XX	03/31/XX
	111.25	2100	Brown Wholesale	421923	02/24/XX	03/31/XX
	969.30	2100	Brown Wholesale	421924	02/24/XX	03/31/XX
	231.67	2100	Brown Wholesale	421990	02/22/XX	03/31/XX
	100.57	2100	Brown Wholesale	422453	02/22/XX	03/31/XX
	80.80	2100	Brown Wholesale	422562	02/23/XX	03/31/XX
	29.30	2100	Brown Wholesale	422676	02/23/XX	03/31/XX
	92.50	2100	Brown Wholesale	422678	02/24/XX	03/31/XX
	10.73	2100	Brown Wholesale	422681	02/24/XX	03/31/XX
	188.96	2100	Brown Wholesale	422988	02/25/XX	03/31/XX
	440.70	2100	Brown Wholesale	422992	02/25/XX	03/31/XX

GENERAL LEDGER REPORT

Account Number	Expense Amount	Vendor	Vendor Name	Invoice Number	Invoice Date	Liability Date

Acct Total

NAME	**JOB COST MASTER (COMPUTER SCREEN)**
NUMBER	**A-7**
ORIGINATOR	**Cost Accounting**
COPIES	**Master maintained in the computer** **Printouts made as required**
DISTRIBUTION	**General Manager, Estimating, General Accounting, Sales**
SIZE	**Printout is condensed to 8½ × 11 in.**

PURPOSE The Job Cost Master is established in the computer for every job started by the company. The top half of the form is completed when the master is set up. It furnishes the identification of the job and establishes if the job is cost-plus or fixed contract amount.

The job cost information on the lower part of the form is completed automatically by the computer from information entered at the accounts payable and payroll entry stations.

The form may be either printed out by the computer as required or displayed on any computer screen authorized by management.

JOB COST MASTER

1. Job Prefix: _____   2. Job No.: _____   3. Job Suffix: _____

4. Job Name: _____

5. Contractor: _____   8. PO/Contract No.: _____

6. Address: _____   9. Estimator: _____

7. City/State/Zip: _____   10. Complete Y/N: _____

11. Cost plus Y/N: _____   12. Job Type: _____   13. Estimated Cost: _____

14. Contract Amount: _____

	Estimate	Current	Last Month
Labor hours	_____	_____	_____
Labor dollars	_____	_____	_____
Labor adder	_____	_____	_____
PO material	_____	_____	_____
Stock material	_____	_____	_____
Tool rental	_____	_____	_____
Subcontract	_____	_____	_____
Direct job expense	_____	_____	_____
Sales tax	_____	_____	_____
Total cost to date	_____	_____	_____
Total contract amount	_____	_____	_____
Percent complete	_____	_____	_____

NAME	**INVOICE**
NUMBER	**A-8**
ORIGINATOR	**Accounting Department**
COPIES	**5**
DISTRIBUTION	**Customer (3 copies), Sales file, Accounting**
SIZE	**8½ × 11 in.**

PURPOSE   The company name and address are necessary to identify your company to your customer.

The invoice number is needed to identify the account for accounts receivable records and to establish an audit trail. The invoice number is also used to locate the invoice to verify billing or to answer possible questions.

The invoice date:

✓ Establishes the date of billing and the date on which the terms of payment start.

✓ Determines the period applicable to this billing.

✓ Allows the computer to age the accounts receivable.

The terms of payment, as noted on the invoice, informs the customer of the dates by which payment must be received for the customer to be entitled to applicable discounts and stay out of arrears.

The customer's name, code number, and address are required not only for mailing but also for setting up your company's account receivable records. The accounts receivable posting becomes an automatic operation each time billing invoices are entered.

The physical location of the job and the customer job number aid your customer in identifying the job for which you are billing.

Form A-8 is an example of a popular style of invoice.

INVOICE

ABC Electric Co.
1410 SOME ROAD, NEW TOWN, MI 19210
Phone (111) 234-5678

INVOICE NO: *96765*

Date: *01/03/93*   Terms: *Net 30 days*

Customer No.: *8096*
Bill to: *Some General Contractor*   Job No.: *123456*
245 Any Street
Home Town, MI 19235

Job Name: *Home Town Airport*	Our Job No.: *30493*	Location: *23000 Lone Road*

Contract amount	*$3,456,123.00*	
Completed to date, *30%*		$1,036,836.00
Less: Previous billing		987,654.00
Net billing this period		$ 49,182.00
Previous billing not paid		50,000.00
Total amount now due and payable		$ 99,182.00

INVOICE

[Company Name
Address
Phone Number]

INVOICE NO:

Date: Terms:

Customer No.:

Bill to: Job No.:

Job Name: Our Job No.:   Location:

Contract amount

Completed to date,

Less: Previous billing   _____

Net billing this period

Previous billing not paid   _____

Total amount now due and payable

NAME	ACCOUNTS RECEIVABLE JOB MASTER (COMPUTER SCREEN)
NUMBER	A-9
ORIGINATOR	Cost Accounting
COPIES	Retained in the computer Printed as required
DISTRIBUTION	As required
SIZE	8½ × 11 in.
PURPOSE	The information required for the completion of the Accounts Receivable Job Master, such as total (TTL) invoices, last billing date, current amount due, retention amount, and last payment (pmt) date, is applied and/or accrued as required. The Accounts Receivable Job Master then becomes the company's record of the contract amount, the change-order amount, and the total amount of billing to date. All the other information on the Accounts Receivable Job Master has been previously entered when the contract for the job was received.

As the invoices are prepared, the entry of the job prefix and job number causes the computer to access the Accounts Receivable Job Master for the job name, the contract number, and the customer's billing address, applying the information directly to the invoice. Also, the computer will access the job type, invoice type, subcontract of prime contract, tax rate, tax method, and progress analysis information shown on the Accounts Receivable Job Master, and use this information to set up the correct invoice form for that job.

The labor adder, overhead, profit, and retention information shown on the Accounts Receivable Job Master will also be applied to the cost amounts shown on the invoices.

ACCOUNTS RECEIVABLE JOB MASTER

1. Job Prefix: _____   2. Job No.: _____   3. Job Suffix: _____

4. Job Name: _____   5. Contract No.: _____

Bill to: _____   Group: _____   GL Account: _____

_____   Job Type: _____   Invoice Type: _____

_____   Sub/Prime: _____   City: _____

_____   Tax Rate: _____   Tax Method: _____

Final Bill Y/N Prog. Annually Flag Y/N

Base Contract	Change-Order	Change-Order	Current Contract
Amount: _____	Number: _____	Amount: _____	Amount: _____
Labor Adder: _____	TTL Invoices: _____	Last Bill date: ____ / ____ / ____	
Overhead: _____	Current Due: _____	Last Payment Date: ____ / ____ / ____	
Profit: _____	Retention: _____		

NAME **SALES JOURNAL**

NUMBER **A-10**

ORIGINATOR **Accounting Department**

COPIES **Retained in computer**
Printed as required

DISTRIBUTION **Available at all computer terminals as authorized**

SIZE **8½ × 14 in. (Copy form at 127% of original.)**

PURPOSE The Sales Journal is the source of information for regular income entries in the General Ledger. As with the Accounts Payable Invoice Register, the information from the Sales Journal may be posted automatically to the income section of the General Ledger through an interface program. Alternatively, the sales figures may be picked up from the journal by a clerk and posted as separate items to the General Ledger accounts.

The accounts receivable entries to customers' job accounts should be made automatically from the Sales Journal by the computer. This eliminates the opportunity for error, which is always present when a clerk is required to transfer information from one journal to another. The invoices from the Sales Journal will be entered by date and invoice number in the correct dollar amount to each customer's account in accounts receivable.

SALES JOURNAL

Invoice Number	Job Number	Job Name	Invoice Amount	Invoice Date
9999	P 99698	Mos 5 Mod 2E probe	-9,560.33	08/26/XX
85710	P 81098	CS-1 demolition	-6,589.18	08/17/XX
85786	P 1110991	W.L. Gore	-8,488.00	08/08/XX
85788	P 1010898 A	POD 2	154.75	08/08/XX
85799	P 1813398	Phoenix Open	69,911.66	08/09/XX
85799	P 1813398	Phoenix Open	-69,911.66	08/09/XX
85801	P 1813398	Phoenix Open	69,778.98	08/09/XX
85534	P 81393	Bldg. 100 proc. mats	19,786.07	08/16/XX
85536	P 1393	Solvent reclaim line	1,478.68	08/17/XX
85537	P 1010898 D	Bldg. 100 chiller	7,660.54	08/17/XX
85538	P 1118098	Mod 2 981 corridor	723.25	08/17/XX
85540	P 81098	CS-1 demolition	7,087.67	08/17/XX
85544	P 81498	East wpws 90,93,9R4	4,861.76	08/17/XX
85545	P 85893	CS-1 ASHE repair	4,556.05	08/17/XX
85546	P 88698	Power for Haworth	1,158.50	08/17/XX
85547	P 55398	CS-1 phase 18	4,548.64	08/17/XX
85554	P 81098 E	Relocate panels CS-1	9,569.55	08/17/XX
85555	P 81098 C	Grounding new column	933.74	08/17/XX
85557	P 1498	Mos 5 MCM up phase	23,611.13	08/17/XX
85558	P 1593	Mos 6 MCM up PR 2	6,351.64	08/18/XX
85559	P 2793	Heat pulse 4100 LD	4,006.41	08/18/XX
85560	P 2193	SPS misc	2,333.18	08/18/XX
85561	P 2993	SPI gas cabinet ser.	3,113.16	08/18/XX
85562	P 21193	Bldg. 101 renov. up	4,447.25	08/18/XX
85563	P 81893	Bldg. 101 renov. up	2,744.84	08/18/XX
85564	P 81493	Hsg. training bldg 90	5,967.73	08/18/XX
85565	P 82093	Mos 5. mod 2E valac.	10,086.90	08/18/XX
85566	P 82493	SPS misc.	3,911.56	08/18/XX
85567	P 82593	DP air mod 15-1	3,256.46	08/18/XX
85568	P 32793	Mod 18/14/ lied resm	6,100.94	08/18/XX
85569	P 33093	SPI bay 1 wafer	1,323.05	08/18/XX
85570	P 43793	SPS bay 25 MCRC 647	194.90	08/18/XX
85571	P 44398	933/94 upgrade oven	3,480.15	08/18/XX
85572	P 99498	Mos 5 mod 2E probe	3,560.33	08/18/XX
85573	P 1111498	Bldg. 100 renov. fdr.	156.45	08/18/XX
85574	P 1118198	Mos 5 WCA monitor	977.63	08/18/XX
85575	P 1118898	Mos 5 furnace	960.95	08/18/XX
85576	P 1813698	Mod 4 mask throw	1,998.76	08/18/XX
85577	P 1814498	SPS bay 25 mods	1,568.33	08/18/XX
85578	P 1814598	Mod 1/2 office	1,257.89	08/18/XX
85580	P 81798	Mtg. discrepancy	3,387.31	08/23/XX
85581	P 76798	Desert Sam Emergency	25,189.30	08/23/XX
85582	P 76598	D. Sam radiation sw.	3,320.79	08/23/XX
85583	P 77798	St. Lukes int lal	2,330.79	08/23/XX
85584	P 1010898	Scottsdale WTP	16,090.70	08/23/XX
85908	P 81398 A	PKWTP devices	5,323.32	08/24/XX
85906	P 54898	Squaw Peak WTP	65,685.00	08/24/XX
85906	P 99898	ASK ind cooling tews	7,032.60	08/24/XX
85907	P 99798	WMOG ltg. modifict.	15,521.00	08/24/XX
85909	P 81593	Kachak field ltg.	14,774.90	08/24/XX
85910	P 88193	WMOG UPS system	7,356.14	08/24/XX
85981	P 82893	Equipment change out	15,316.97	08/25/XX
85985	P 99698	Mos 5 mod 2E probe	3,476.95	08/26/XX
85986	P 1893	Tanner Chapel Manor	17,325.00	08/30/XX
85954	P 1110891	23rd Ave.-Phase 91	99,065.00	08/30/XX
85956	P 45891	23rd Ave. Waste Water	45,405.15	08/30/XX
		Total	557,371.25	

Large contracts total	294,998.17
Motorola total	91,994.19
Small contracts total	200,375.29

SALES JOURNAL

Invoice Number	Job Number	Job Name	Invoice Amount	Invoice Date

Total

Large contracts total
Small contracts total

NAME	**TIME AND MATERIAL BILLING REGISTER**
NUMBER	**A-11**
ORIGINATOR	**Accounting Department**
COPIES	**Printed out by the computer, using information from invoices**
DISTRIBUTION	**Accounting Department**
SIZE	**11 × 8½ in.**

PURPOSE

The Time and Material Billing Register differs from the large project Sales Journal in the additional detail required when billing time and material or cost-plus work. Form A-11 illustrates the detail required such as labor cost, labor adder, material cost, sales tax, and, when required, the labor rate per hour being billed.

The register may also calculate the gross profit amount for each billing and the percentage of profit earned. This provides management with a quick check on the efficiency of the service department.

TIME AND MATERIAL BILLING REGISTER

Job Number	Job Name	Invoice	TY	Date	Invoice Amount	Labor Cost	Labor Addr	Material Cost	Sales Tax	Gross Profit	Gross Profit %	Labor Rate
7 5390	135 E. Orien. Tempe	86018	76	04/09/XX	586.77	251.75	106.09	80.30	23.79	117.94	20.10	17.25
7 1165	Ag Dept. of Trans.	86080	79	04/13/XX	394.55	215.26	55.26	15.09	16.55	55.39	14.72	.00
7 3660	Motorola Mesa	86081	75	04/13/XX	300.17	95.50	39.25	94.58	18.17	55.10	19.36	.00
7 3660	Motorola Mesa	86082	75	04/13/XX	326.52	150.75	61.51	25.01	13.22	72.43	22.20	.00
7 3660	Motorola Mesa	86083	75	04/13/XX	773.25	.00	.00	645.12	31.35	96.55	12.53	.00
7 3660	Motorola Mesa	86084	75	04/13/XX	1,166.25	591.35	242.47	.00	47.25	285.15	24.45	.00
7 3660	Motorola Mesa	86085	75	04/13/XX	229.30	95.50	39.25	35.53	9.30	49.09	21.41	.00
7 3660	Motorola Mesa	86086	75	04/13/XX	1,362.22	509.40	205.55	355.39	55.22	236.96	17.35	.00
7 3660	Motorola Mesa	86087	75	04/13/XX	1,585.39	395.50	162.16	656.53	64.39	309.51	19.49	.00
7 3660	Motorola Mesa	86088	75	04/13/XX	1,317.40	271.76	111.48	597.47	53.40	283.35	21.51	.00
7 1860	91 st Ave WWTP	86090	82	04/13/XX	280.00	166.00	68.06	.00	11.69	34.25	12.23	20.75
7 3290	Lescher & Mahoney	86091	75	04/13/XX	36.52	16.00	6.56	.00	1.52	12.44	34.06	16.00
7 4350 8		86092	75	04/13/XX	216.10	78.75	32.29	87.59	9.02	65.15	31.54	15.75
7 4680	La Posada	86093	82	04/13/XX	1,640.00	404.50	165.55	353.05	65.44	645.16	39.52	20.22
7 4670		86094	75	04/13/XX	109.57	64.05	26.26	.00	4.57	14.69	13.41	21.35
7 5045	Thunderbirds	86095	76	04/13/XX	184.91	55.40	35.01	53.67	7.72	23.11	12.50	21.35
7 5390 A	135 E. Orien. Tempe	86096	76	04/13/XX	935.03	.00	.00	757.00	35.03	113.00	12.05	.00
				Grand Totals	59,182.73	17,750.30	7,277.71	19,235.95	2,351.19	11,876.55	20.09	
Motorola total						41,014.65	12,673.12	5,196.03	13,931.42	1,662.59	7,551.49	18.41
Small contracts total						18,105.08	5,077.18	2,081.65	5,904.56	719.30	4,325.36	23.59
Large contracts total						.00	.00	.00	.00	.00	.00	.00

	# of Jobs	Invoice Amount	Gross Profit	GP %
Inv. type 5	10	1,253.89	467.53	25.88
Inv. type 6	9	13,597.19	2,643.25	19.44
Motorola	44	41,014.65	7,551.49	18.41
Base Quotes	6	2,657.00	1,214.55	45.71

TIME AND MATERIAL BILLING REGISTER

Job Number	Job Name	Invoice	TY	Date	Invoice Amount	Labor Cost	Labor Addr	Material Cost	Sales Tax	Gross Profit	Gross Profit %	Labor Rate
				Grand Totals								

Small contracts total
Large contracts total

# of Jobs	Invoice Amount	Gross Profit	GP %

NAME **ACCOUNTS RECEIVABLE AGING REPORT**

NUMBER **A-12**

ORIGINATOR **Accounting Department**

COPIES **2**

DISTRIBUTION **Accounts Receivable, collection file**

SIZE **8½ × 11 in.**

PURPOSE Every receivable account must be aged each month according to due date, indicating the amounts due and past due in the following categories:

Current—under 30 days

30 days past due

60 days past due

90 days past due

The condition of the current asset known as accounts receivable is determined by the aging record. Generally, any account that is 90 days or more past due is in perilous condition.

The Accounts Receivable Aging Report is readily available each month immediately after the Sales Journal and Cash Receipts Journal (see Form A-13) have been posted to the computer. The aging report is furnished automatically by the computer without the need to enter further data. Close control of all accounts receivable amounts enhances available working capital.

ACCOUNTS RECEIVABLE AGING REPORT

Date	Invoice Number	Balance	Current	30 Days	60 Days	90 Days
P 1193	U.S. West Communications		DDA system 1st floor			
08/23/XX	25721	18.06			18.06	
Job Total		18.06	.00	.00	18.06	.00
P 1393	Motorola, Inc.		Solvent reclaim ltng			
08/17/XX	25836	1,478.48			1,478.48	
Job Total		1,478.48	.00	.00	1,478.48	.00
P 1493	Motorola, Inc.		Mes 5 MLM exp phase			
03/18/XX	25857	23,611.13		23,611.13		
Job Total		23,611.13	.00	23,611.13	.00	.00
P 1593	Motorola, Inc.		Mes 6 MLM exp ph 2			
03/18/XX	25858	6,251.64		6,251.64		
Job Total		6,251.64	.00	6,251.64	.00	.00
P 2793	Motorola, Inc.		Heat Pulse 4100 LO			
03/18/XX	25859	4,006.41		4,006.41		
Job Total		4,006.41	.00	4,006.41	.00	.00
P 2893	Motorola, Inc.		8P misc			
2/22/XX	25708	4,549.42			4,549.42	
03/18/XX	25860	2,223.12		2,223.12		
Job Total		6,772.54	.00	2,223.12	4,549.42	.00
P 2993	Motorola, Inc.		8P1 gas cabinet rev.			
03/18/XX	25861	3,112.16		3,112.16		
Job Total		3,112.16	.00	3,112.16	.00	.00
P 21093	McCarthy		CS-1 demolition			
03/17/XX	25840	7,087.67		7,087.67		
Job Total		7,087.67	.00	7,087.67	.00	.00
P 21093B	McCarthy		Relocate panels CS-1			
03/17/XX	25854	9,569.55		9,569.55		
Job Total		9,569.55	.00	9,569.55	.00	.00
P 21093C	McCarthy		Grounding new column			
03/17/XX	25855	983.74	983.74			
Job Total		983.74	.00	983.74	.00	.00

ACCOUNTS RECEIVABLE AGING REPORT

Date	Invoice Number	Balance	Current	30 Days	60 Days	90 Days

	NAME	**CASH RECEIPTS JOURNAL**
	NUMBER	**A-13**
	ORIGINATOR	Accounting Department
	COPIES	2
	DISTRIBUTION	Comptroller, Accounting file
	SIZE	8½ × 11 in.

PURPOSE   The Cash Receipts Journal enables a company's bookkeeper to record all monies received by the company and apply the receipts to the correct General Ledger and accounts receivable customer accounts.

Bank deposit receipts are generally attached to the Cash Receipts Journal for a record of the deposit.

CASH RECEIPTS JOURNAL

Job Number	Job Name	Date Paid	Invoice Number	Amount	Routing Slip	GL Account	Check Total
P 1213392	Phoenix Open	04/13/XX	25801	69,772.92	91-2	111 0	69,772.92
P 1111492	Bldg. 100 emerg. fdr.	04/13/XX	25688	205.32	70-1558	111 0	
P 1493	Mos 5 MLM exp phase	04/13/XX	25727	5,655.25	70-1558	111 0	
7 3660	Motorola Mesa	04/13/XX	25757	803.57	70-1558	110 0	
7 3660	Motorola Mesa	04/13/XX	25830	514.87	70-1558	110 0	7,179.01
P 1010092	Scottsdale WTP	04/13/XX	25704	6,444.00	40-2	111 0	6,444.00
P 1112892	Regenerative blower	04/13/XX	25322	5,245.00	91-265	111 0	5,245.00
P 1010092	Scottsdale WTP	04/13/XX	25704	716.00	40-2	111 0	716.00
7 3660	Motorola Mesa	04/13/XX	25968	207.13	70-1558	110 0	207.13
			Total	89,564.06			
111 0	88,038.49						
110 0	1,525.57						

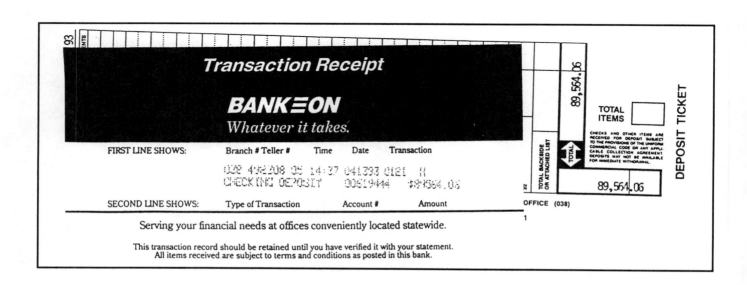

CASH RECEIPTS JOURNAL

Job Number	Job Name	Date Paid	Invoice Number	Amount	Routing Slip	GL Account	Check Total

NAME	**SCHEDULE OF FIXED ASSETS**
NUMBER	**A-14**
ORIGINATOR	**Accounting Department**
COPIES	**3**
DISTRIBUTION	**General Manager, Accounting, Tool Department**
SIZE	**14 × 11 in.**

PURPOSE Tools, equipment, automobiles, trucks, and furnishings represent a sizable investment for any company. It is imperative to have a record of these assets to enable management to determine when additions or replacements are required and to maintain a current inventory of the assets.

Most of the fixed assets are subject to depreciation for tax and replacement purposes. The recurring calculation-of-depreciation nightmare experienced by the accounting departments of most businesses has been solved by the fixed assets computer software now available. All these programs allow you to make one entry for each fixed asset, noting the asset class, the cost, depreciable amount, and type of depreciation applicable to the asset, and accrued depreciation to date.

From that point, a simple update procedure is performed each accounting period, and the computer not only calculates the current depreciation for all assets listed, but also lists the total depreciation to date by asset. The computer printout is then used to back up General Ledger depreciation amounts, both for book and tax purposes.

Form A-14 illustrates a typical Schedule of Fixed Assets and depreciation expense report, which is created automatically by the computer at the end of an accounting period. The computer is capable of using this information to print a simple list of the fixed assets showing only the asset number, description, date acquired, and the acquisition value.

SCHEDULE OF FIXED ASSETS

Date: 3/01/XX

Asset Class	Asset Number	Asset Description	Date Acquired	Acquisition Value	Depr. Method	Life, Year Month	B/salvage L sect. 179	Depreciable Basis	Last Depreciation	Prior Total Accumulation	Depr. This Run	Current YTD Depreciation	Total Accumulated Depreciation
1	1200	Building #1	3/8/88	$75,000	SL	30		$75,000.00	$2,500.00	$5,000.00	$2,500.00	$2,500.00	7,500.00
3	234	Chev P/U	3/1/90	12,000	DDB	4	B	10,000.00			5,000.00	5,000.00	5,000.00
4	555	Bender	1/1/91	6,000	SL	3	B	6,000.00		334.00	334.00	334.00	
2	2415	Desk	3/1/88	250	SL	10		250.00	50.00	50.00	25.00	25.00	75.00
Total				$93,250.00				$93,250.00	$2,550.00	$5,050.00	$7,859.00	$7,859.00	$12,909.00

SCHEDULE OF FIXED ASSETS

Date:

Asset Class	Asset Number	Asset Description	Date Acquired	Acquisition Value	Depr. Method	Life, Year Month	B/salvage L sect.179	Depreciable Basis	Last Depreciation	Prior Total Accumulation	Depr. This Run	Current YTD Depreciation	Total Accumulated Depreciation
1													
3													
4													
2													
Total													

NAME	**GENERAL LEDGER (COMPUTER SCREEN)**
NUMBER	**A-15**
ORIGINATOR	**Accounting Department**
COPIES	**Retained in computer**
DISTRIBUTION	**Not distributed; used to make computer entries**
SIZE	**8½ × 11 in.**

PURPOSE The computerized General Ledger is set up in the same manner as in a manual system and divided according to the four basic categories: assets, liabilities, income, and expense. It is necessary to have a General Ledger Chart of Accounts (see Form A-16), which, of course, is expandable, and a series of logical account codes. One primary difference in the account codes between the manual system and the computerized system is that the computer system requires a decimal series of code numbers, while the manual mode allows an alphanumeric system. For example, the fixed asset account in the computer system would appear as:

180.0 BUILDINGS

180.1 ALLOWANCE FOR DEPRECIATION—BUILDINGS

In the manual system the same accounts might appear as:

180 BUILDINGS

180R RESERVE FOR DEPRECIATION—BUILDINGS

One reason for this difference in account coding is that the computer is more easily programmed to recognize numerical fields than to read and process alphanumeric information. Another is easier input from a ten-key device. The input screen for the General Ledger is shown as Form A-15.

GENERAL LEDGER

Account Distribution

Journal Entry Number	Date	Description	Acct. Number	Account Title	Debit	Credit
1	3/01/XX	A/P	201	A/P	3,000.00	
			101	Cash in bank		3,000.00
2	3/01/XX	A/R	101	Cash in bank	2,500.00	
			110	A/R		2,500.00
3	3/01/XX	Sales	110	A/R	15,250.00	
			302	Income		15,250.00

Display   Next   Previous   Enter   Change   Remove   Help   Quit

GENERAL LEDGER

Journal Entry Number	Date	Description	Acct. Number	Account Title	Debit	Credit

Account Distribution

Display   **Next**   **Previous**   **Enter**   **Change**   **Remove**   **Help**   **Quit**

NAME	**GENERAL LEDGER CHART OF ACCOUNTS**
NUMBER	**A-16**
ORIGINATOR	**Chief Accountant**
COPIES	**All Accounting personnel**
DISTRIBUTION	**All Accounting and Cost Accounting personnel**
SIZE	**8½ × 11 in.**

PURPOSE The published General Ledger Chart of Accounts establishes the basis for a standard companywide accounting system. This system enables every person responsible for entering items in the accounts of the company to properly code and therefore segregate the various charges and credits.

Most of the entries made to the General Ledger originate from the various accounting journals such as:

Cash in bank: From the Cash Received Journal.

Accounts receivable: From the Sales Journal in total amounts.

Fixed assets: From the Accounts Payable Invoice Register.

Liabilities: Credit entries from the Accounts Payable Invoice Register and debits from the Check Register.

Income: From the Sales Journal.

Expense: From the Accounts Payable Invoice Register.

GENERAL LEDGER CHART OF ACCOUNTS
Assets

Account Number	Account Name
	CURRENT ASSETS
100	Cash on Hand
101	Cash in Bank
102	Certificates of Deposit
105	Accounts Receivable, Regular
110	Accounts Receivable, Progress
111	Accounts Receivable, Retail Sales
115	Notes Receivable, Current
120	Inventory Stock
120.1	Inventory Progress
140	Unbilled Amounts on Completed Contracts
140.1	Costs and Estimated Earnings on Contracts in Progress in Excess of Billings
150	Prepaid Expenses and Other current Assets
	OTHER ASSETS
151	Land Held for Investment
152	Cash Surrender Value of Life Insurance
153	Investments (Stocks and Bonds)
	FIXED ASSETS
160	Land
161	Buildings and Improvements
162	Office Equipment
163	Furniture and Fixtures
164	Autos and Trucks
165	Construction Equipment
166	Tools

GENERAL LEDGER CHART OF ACCOUNTS
Liability Accounts

Account Number	Account Name
	CURRENT LIABILITIES
205	Accounts Payable
206	Retention
208	Accrued Payroll
209	Payroll Taxes
210	Union Benefits
211	Amounts Withheld from Employees
215	Additional Costs on Completed Contracts
216	Amounts Billed in Excess of Costs and Estimated Earnings on Contracts in Progress
218	Current Income Taxes
219	Deferred Income Taxes Owed
220	Current Portion of Long-Term Debt
221	Other Taxes Due, Current Period
	OTHER LIABILITIES
225	Mortgages
226	Notes Payable Later Than 1 Year
	STOCKHOLDERS EQUITY AND NET WORTH
260	Common Stock
261	Additional Paid-In Capital
262	Retained Earnings

GENERAL LEDGER CHART OF ACCOUNTS
Income Accounts

Account Number	Account Name
	INCOME
301	Income, Service Jobs
302	Income, Progress Jobs
303	Income, Retail Sales
310	Interest Income
315	Discounts Earned
329	Income, Disposition of Assets
340	Other Miscellaneous Income

GENERAL LEDGER CHART OF ACCOUNTS
Expense Accounts

Account Number	Account Name
	EXPENSE
501	Advertising and Promotion
502	Automobiles and Trucks
503	Bad Debts
504	Charitable Contributions
505	Collection Expense
506	Depreciation and Amortization
507	Dues and Subscriptions
508	Education Expenses
509	Overhead Employee Benefits
510	Freight and Express
511	Heat, Light, Power, and Water
512	Company Insurance—General
514	Overhead, Employee Workmens Compensation
515	Public and Employers Liability Insurance
516	Legal and Accounting
518	Miscellaneous
520	Office Supplies and Expense
521	Pension and Profit-Sharing Plans
522	Rent
523	Repairs and Maintenance
525	Plan, Bid Bonds, Estimating, and Engineering Expense
526	Salaries and Wages—Overhead Personnel
528	Shop Supplies and Expense
530	Small Tools
532	Taxes and Licenses—General
534	Taxes—Overhead Payroll
536	Telephone, Telegraph, and Postage
537	Travel—Administrative
550	Other Expenses

NAME	**WEEKLY PAYROLL (COMPUTER SCREEN)**
NUMBER	**A-17**
ORIGINATOR	**Payroll Department**
COPIES	**Retained in computer, not distributed**
DISTRIBUTION	**Not distributed**
SIZE	**8½ × 5½ in.**

PURPOSE   The computer entry screen for Weekly Payroll information is shown in Form A-17. The computer uses the information entered on this screen to produce all the management payroll information reports automatically. The entry screen is interfaced with the Employee Personnel Record (Form M-20) to determine pay rate, pay code, deductions, and taxes applicable to each employee.

When all the entries for the current period payroll have been entered, the computer automatically prints a check for each employee and a payroll check register.

WEEKLY PAYROLL

Employee Number: _____   Job Number: _____   Labor Code: _____   System Number: _____

Div. Number: _____   Regular Hours: _____   OT Hours: _____   DT Hours: _____   Other: _____

Display   **Next**   **Previous**   **Enter**   **Change**   **Remove**   **Help**   **Quit**

PAYROLL REGISTER

NUMBER	**A-18**
ORIGINATOR	Payroll Department
COPIES	I
DISTRIBUTION	Payroll Department
SIZE	14 × 11 in. (Copy form at 125% of original.)

PURPOSE
The Payroll Register is used to balance man-hours paid to actual job time cards, as well as to maintain a written record of the amounts paid to employees. The register also lists all applicable taxes and deductions withheld and is used to balance the various reports.

Because all applicable tax and union reports are handled automatically by the computer, the reports may be assumed to be accurate and therefore do not require any checking time by the staff. Once the payroll register has been balanced, all the resulting reports are correct and require no further audit.

The computer accumulates payroll information by employee from the current payroll and applies it to the individual employee's personal account to provide a permanent record of that employee's earnings.

PAYROLL REGISTER

	Regular	Overtime	Double Time	Other	Gross	OASDI—Medicare	Federal	State	Dues	Deducts	Net
11 David Aberta			04/07/XX 20792								
Hours	40.00	.00	.00			40.92					
Amount	660.00	.00	.00	.00	660.00	9.57	80.86	22.64	13.20	.00	492.81
121 Jon C. Anderson			04/07/XX 20793								
Hours	40.00	.00	.00			37.20					
Amount	600.00	.00	.00	.00	600.00	8.70	58.59	12.89	12.00	.00	470.62
171 Wiley C. Atkins			04/07/XX 20794								
Hours	40.00	.00	.00			39.68					
Amount	640.00	.00	.00	.00	640.00	9.28	64.59	14.21	12.80	.00	499.44
181 Mark A. Avery			04/07/XX 20795								
Hours	40.00	.00	.00			39.68					
Amount	640.00	.00	.00	.00	640.00	9.28	113.22	36.23	12.80	.00	428.79
375 Jamie F. Bergendahl			04/07/XX 20796								
Hours	40.00	.00	.00			25.57					
Amount	412.40	.00	.00	.00	412.40	5.98	47.73	9.55	8.25	.00	315.32
401 Robert E. Blake			04/07/XX 20797								
Hours	40.00	.00	.00			39.56					
Amount	638.00	.00	.00	.00	638.00	9.25	71.07	22.74	12.76	31.90	450.72
641 Robert W. Buttler			04/07/XX 20798								
Hours	36.00	.00	.00			47.65					
Amount	768.60	.00	.00	.00	768.60	11.14	136.57	30.05	15.37	.00	527.82
741 Thomas L. Canizzero			04/07/XX 20799								
Hours	40.00	.00	.00			56.42					
Amount	910.00	.00	.00	.00	910.00	13.20	77.98	24.95	18.20	.00	719.25
791 Charles J. Celigno			04/07/XX 20800								
Hours	40.00	.00	.00			31.62					
Amount	510.00	.00	.00	.00	510.00	7.40	48.81	15.62	10.20	25.50	370.85
951 Patrick N. Compers			04/07/XX 20801								
Hours	38.00	.00	.00			37.70					
Amount	608.00	.00	.00	.00	608.00	8.82	53.01	11.66	12.16	.00	484.65
1111 Jack L. Dickson			04/07/XX 20802								
Hours	40.00	.00	.00			40.30					
Amount	650.00	.00	.00	.00	650.00	9.57	8.70	22.50	13.10	.00	555.83

PAYROLL REGISTER

	Regular	Overtime	Double Time	Other	Gross	OASDI— Medicare	Federal	State	Dues	Deducts	Net
Hours											
Amount											
Hours											
Amount											
Hours											
Amount											
Hours											
Amount											
Hours											
Amount											
Hours											
Amount											
Hours											
Amount											
Hours											
Amount											
Hours											
Amount											
Hours											
Amount											

CHAPTER THREE
FINANCIAL

FORM NO.	FORM NAME	PAGE
F-1	Balance Sheet	86
F-2	Statement of Profit and Loss	88
F-3	Summary of Income and Expense	88
F-4	Optimal Values for Critical Financial Ratios and Percentages	90
F-5	Working Capital Requirement	92

NAME **BALANCE SHEET**

NUMBER **F-1**

ORIGINATOR **Accounting Department**

COPIES **2**

DISTRIBUTION **General Manager, Chief Accountant**

SIZE **8½ × 11 in.**

PURPOSE Once all the entries for the accounting period have been made to the General Ledger, the computer sorts, adjusts, and accumulates each account to the current balance.

Then a trial balance is obtained to ascertain that the General Ledger is in balance. At this point any adjustments, such as current earnings or loss, should be entered on the proper accounts in the General Ledger through the General Journal.

The Balance Sheet is the overall picture of the current financial condition of the business. It is imperative that the general manager or someone on the GM's staff be capable of reading and thoroughly understanding the Balance Sheet, the Statement of Profit and Loss (Form F-2), and the ratios developed from the information shown on these statements.

ABC Electric Co.
BALANCE SHEET
For Period *Any Month End*

Assets		Liabilities	
Current Assets		*Current Liabilities*	
Cash	$50,000	Accounts payable, including retention of $10,000	$95,000
Certificates of deposit	165,000		
Accounts receivable	75,000	Accrued payroll, payroll taxes, and benefits	2,000
Unbilled amounts on completed contracts	10,000	Addition costs on completed contracts	2,500
Material and supplies inventory	20,000	Amounts billed in excess of cost and estimated earnings on contracts in progress	54,500
Costs and estimated earnings on contracts in progress in excess of billings	40,000	Other current liabilities	5,000
		Income taxes	
Prepaid expense and other current assets	5,000	Current	4,500
		Deferred	10,000
Total current assets	$365,000	Current portion of long-term debt	6,500
Other Assets		Total current liabilities	$180,000
Land held for investment	$75,000	*Long-Term Debt*	$ 80,000
Cash surrender value of life insurance	16,814	**Stockholders Equity—Net Worth**	
Total other assets	$91,814	Common stock, par value $10.00 per share, authorized 10,000 shares, issued and outstanding 2,000 shares	$ 10,000
Fixed Assets			
Land	15,000	Additional paid-in capital	20,000
Building and improvements	65,000	Retained earnings	268,814
Construction equipment	75,000		$298,814
Vehicles	3,000		
Office equipment	4,000		
Total fixed assets	$162,000		
Less: Accumulated depreciation	60,000		
	102,000		
Total assets	$558,814	*Total liabilities and net worth*	$558,814

BALANCE SHEET
For Period

Assets

Current Assets
- Cash
- Certificates of deposit
- Accounts receivable
- Unbilled amounts on completed contracts
- Material and supplies inventory
- Costs and estimated earnings on contracts in progress in excess of billings
- Prepaid expense and other current assets
- Total current assets

Other Assets
- Land held for investment
- Cash surrender value of life insurance
- Total other assets

Fixed Assets
- Land
- Building and improvements
- Construction equipment
- Vehicles
- Office equipment
- Total fixed assets
- Less: Accumulated depreciation

Total assets

Liabilities

Current Liabilities
- Accounts payable, including retention of $10,000
- Accrued payroll, payroll taxes, and benefits
- Addition costs on completed contracts
- Amounts billed in excess of cost and estimated earnings on contracts in progress
- Other current liabilities
- Income taxes
  - Current
  - Deferred
- Current portion of long-term debt
- Total current liabilities

Long-Term Debt

Stockholders Equity—Net Worth

Common stock, par value $10.00 per share, authorized, 10,000 shares, issued and outstanding 2,000 shares
Additional paid-in capital
Retained earnings

Total liabilities and net worth

NAMES
NUMBERS
ORIGINATOR
COPIES
DISTRIBUTION
SIZE
PURPOSE

STATEMENT OF PROFIT AND LOSS
SUMMARY OF INCOME AND EXPENSE

NAMES STATEMENT OF PROFIT AND LOSS / SUMMARY OF INCOME AND EXPENSE

NUMBERS F-2 / F-3

ORIGINATOR Chief Accountant

COPIES 2 of each

DISTRIBUTION General Manager, Chief Accountant

SIZE 8½ × 11 in.

PURPOSE The Statement of Profit and Loss (Form F-2) must be issued every month for proper control. Commonly referred to as the P&L, or sometimes income statement, it is then part of the financial statement for the current accounting period. This timely report allows management to make comparisons to the budget. In most cases the bottom line, or profit figure, is of primary importance, but the sales figure (top line) is the controlling figure. All other financial considerations are dependent on the sales figure because, without achieving budgeted sales, it is impossible to control the other expense items and make a profit.

In addition to the P&L, a detailed Summary of Income and Expense (Form F-3) listing all the elements of cost, should be prepared, with each element represented as a percentage of sales. The percentages shown on Form F-3 are all based on the sales figure. The percentage adders used in pricing estimates, material, or labor are based on prime cost. So these adders have to be recalculated as percentages of cost to enable the company to recover overhead and earn a profit.

ABC Electric
STATEMENT OF PROFIT AND LOSS
For Period Ending ____ / ____ / ____

		Percentage of Sales (%)
Sales	$2,600,000	100.00
Cost of sales:		
Material	$936,000	36.00
Labor	728,000	28.00
Payroll tax	214,240	8.24
Direct job expense	104,000	4.00
Subcontract	91,000	3.50
Prime cost	$2,073,240	79.70
Gross profit	526,760	20.26
Overhead costs	473,200	18.20
Operating profit	$ 53,560	2.06
Income taxes	18,746	.72
Net profit	$ 43,814	1.34
Retained earnings, beginning of year	$ 225,000	
Retained earnings to date	$ 268,814	

SUMMARY OF INCOME AND EXPENSE

	Group 1: Annual Direct Payroll under $250,000		Group 2: Annual Direct Payroll $250,000 to $500,000		Group 3: Annual Direct Payroll $500,000 to $1,000,000		Group 4: Annual Direct Payroll $1,000,000 to $2,500,000	
	Average Dollars	Percentage of Sales	Average Dollars	Percentage of Sales	Average Dollars	Percentage of Sales	Average Dollars	Percentage of Sales
Total sales	711,556	100.00	1,476,511	100.00	2,830,161	100.00	6,042,106	100.00
Prime cost:								
Material	258,676	36.34	569,621	38.57	1,064,257	37.60	2,264,090	37.47
Direct labor wages	149,533	21.01	361,549	24.48	715,648	25.29	1,647,695	27.27
Labor adder	49,141	6.90	108,583	7.35	259,290	9.16	571,538	9.46
Other direct job expense	16,895	2.37	47,069	3.19	77,914	2.75	211,784	3.68
Subcontract expense	29,532	4.15	47,911	3.24	116,429	4.12	206,399	3.42
Total prime cost	503,777	70.77	1,134,734	76.84	2,233,537	78.92	4,901,464	81.24
Gross income	207,079	29.23	342,077	23.16	596,684	21.08	1,133,642	18.76
Total overhead expense	175,471	24.65	286,506	19.40	490,369	17.33	922,989	15.26
Operating profit	32,608	4.58	55,571	3.76	106,255	3.75	210,653	3.49
Interest expense	5,094	.71	9,883	.67	13,759	.49	30,665	.51
Nonoperating expense	5,514	.77	7,509	.51	18,598	.66	39,690	.66
Net profit before tax	33,078	4.65	53,194	3.60	111,098	3.93	219,678	3.64
Income taxes	9,278	1.30	13,180	.89	33,880	1.20	60,697	1.00
Net profit after tax	23,800	3.34	40,014	2.71	77,218	2.73	159,981	2.63

STATEMENT OF PROFIT AND LOSS
For Period Ending ____ / ____ / ____

		Percentage of Sales (%)
Sales	_____	_____
Cost of sales:		
Material	_____	_____
Labor	_____	_____
Payroll tax	_____	_____
Direct job expense	_____	_____
Subcontract	_____	_____
Prime cost	_____	_____
Gross profit	_____	_____
Overhead costs	_____	_____
Operating profit	_____	_____
Income taxes	_____	_____
Net profit	_____	_____
Retained earnings, beginning of year	_____	
Retained earnings to date	_____	

SUMMARY OF INCOME AND EXPENSE

	Group 1: Annual Direct Payroll under $250,000		Group 2: Annual Direct Payroll $250,000 to $500,000		Group 3: Annual Direct Payroll $500,000 to $1,000,000		Group 4: Annual Direct Payroll $1,000,000 to $2,500,000	
	Average Dollars	Percentage of Sales	Average Dollars	Percentage of Sales	Average Dollars	Percentage of Sales	Average Dollars	Percentage of Sales
Total sales								
Prime cost:								
Material								
Direct labor wages								
Labor adder								
Other direct job expense								
Subcontract expense								
Total prime cost	____	____	____	____	____	____	____	____
Gross income								
Total overhead expense	____	____	____	____	____	____	____	____
Operating profit								
Interest expense								
Nonoperating expense	____	____	____	____	____	____	____	____
Net profit before tax								
Income taxes	____	____	____	____	____	____	____	____
Net profit after tax								

NAME	**OPTIMAL VALUES FOR CRITICAL FINANCIAL RATIOS AND PERCENTAGES**
NUMBER	**F-4**
ORIGINATOR	**Accounting Department**
COPIES	**2**
DISTRIBUTION	**General Manager, Chief Accountant**
SIZE	**8½ × 11 in.**
PURPOSE	Financial ratios are easily defined but often misunderstood. Simply, they are comparisons of any two figures taken from financial statements. They may be expressed as either ratios or percentages. Financial ratios are often used without identification. For instance, it is common to see income statements prepared with a column dedicated to expressing each line item as a percentage of sales.

Financial ratios should be viewed with caution: They are only one type of many measurements of corporate or company health. Individually, they reflect only strength or weakness in a particular area, not in the overall picture. Furthermore, the definition of a "good" ratio may differ from company to company and from industry to industry.

Ratios are meaningless without standards for comparison. Quite often the easiest comparison is to the industry standards published by trade organizations and financial institutions. The National Electrical Contractors Association (NECA) publishes the most useful comparative ratios information for the use in the electrical contracting industry. However, frequently these "standards" are just averages, and they result in comparisons to a theoretical mediocre company.

Form F-4 lists the Optimal Values for Critical Financial Ratios and Percentages for the electrical contracting industry.

OPTIMAL VALUES FOR CRITICAL
FINANCIAL RATIOS AND PERCENTAGES

Ratio	Ideal value
Current ratio	2:1
Fixed assets/net worth	40%
Current liabilities/net worth	80%
Total liabilities/net worth	100%
Gross revenue/working capital	10:1
Gross revenue/net worth	9.6:1
Debt/total capitalization	50%
Long-term liabilities/working capital	25%
Net worth/current assets	63%
Net income/net worth*	25%
Net income/working capital*	30%
Net income/gross revenue*	8%

*Net income before taxes.

NAME	# WORKING CAPITAL REQUIREMENT
NUMBER	**F-5**
ORIGINATOR	**Chief Accountant**
COPIES	**2**
DISTRIBUTION	**General Manager, Chief Accountant**
SIZE	**8½ × 11 in.**
PURPOSE	*Working capital* is the dollar difference obtained by subtracting current liabilities from current assets. For example, working with the current liabilities and current assets from ABC Electric Co.'s Balance Sheet (Form F-1):

Working capital = Current assets − Current liabilities

$$= \$365{,}000 - \$180{,}000$$
$$= \$185.000$$

The turnover of working capital becomes extremely important to the financial health of the business because too quick a turnover can result in the business's not being able to generate enough cash to retire current liabilities and pay current expenses. The turnover of working capital is expressed by the ratio of gross revenue to working capital. The comfortable range in the electrical industry is a ratio of between 6:1 to 8:1. With efficient management and field operations, along with the prompt collection of accounts receivable, the company can operate successfully with a ratio of up to 10:1. A ratio higher than 7:1 is preferred. The formula is

$$\text{Ratio} = \frac{\text{Gross revenue}}{\text{Working capital}}$$

If capital is turning at a low rate, sales may be below normal. Or, if this ratio is low, the company may have more current items than it needs and should be looking for more business or for other investment opportunities.

Note: Gross revenue is the controlling factor, and the preceding formula must be used to enable management to control the turnover of working capital.

Form F-5 illustrates the optimum overall distribution of the elements of the Balance Sheet. The *current ratio* reflects the number of times current assets "cover" current liabilities. In other words, if the company liquidated all current assets and paid off all current liabilities, how much of the assets would be left over? The pie chart in Form F-5 shows the following ratio:

$$\text{Current ratio} = \frac{\text{Current assets}}{\text{Current liabilities}}$$

$$= \frac{80}{40} = 2.1$$

WORKING CAPITAL REQUIREMENT

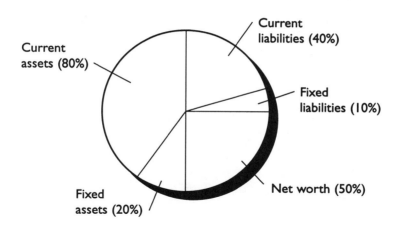

CHAPTER 4
DATA PROCESSING

FORM NO.	FORM NAME	PAGE
D-1	Pricing Database for Pricing and Extensions of the Estimate (Computer Screen)	96
D-2	Pricing Database for Estimate Summary (Computerized)	98
D-3	Weekly Payroll Report	100
D-4	Progress Analysis	102

NAME	**PRICING DATABASE FOR PRICING AND EXTENSIONS OF THE ESTIMATE (COMPUTER SCREEN)**
NUMBER	**D-1**
ORIGINATOR	**Estimating Clerk**
COPIES	**2**
DISTRIBUTION	**Estimator, Computer Department**
SIZE	**8½ × 11 in.**
PURPOSE	The computer establishes a standardized, orderly, efficient, and accurate means of estimating and requires a minimal amount of entries to generate several valuable reports. These include a detailed Bill of Material by divisions of the estimate, and a consolidated Bill of Material for the entire estimate, listing all the material by type, which can be used in obtaining quotes from vendors and checking on-hand inventory to determine what material does not have to be purchased. Total labor requirements, as well as purchasing budget guidelines, become available from the information generated in the basic estimate.

Form D-1 illustrates a typical entry screen necessary to access the pricing data base.

**PRICING DATABASE FOR PRICING
AND EXTENSIONS OF THE ESTIMATE**

Estimate Number: *102* Job Number: *P31492* System: *3* System Name: _____

Branch: _____

Section Number: _____ Estimated by: *RJJ* Date: *01/03/XX*

Job Name: *J. Warehouse* Column: *A*

Code Number	Quantity	Unit	Factor (%)
10001	200	100/ft	+10
10002	100	100/ft	+10
40015	25	ea.	
40025	50	ea.	
50030	440	100/ft	
50035	210	100/ft	

Display	Next	Previous	Enter	Change	Remove	Help	Quit

Code Number	Description	Unit	Material Price			Labor Units		
			A	B	C	A	B	C
10001	½" EMT	100/ft	$50.00	$55.00	$60.00	4.0	4.4	5.0
10002	¾" EMT	100/ft	60.00	66.00	72.00	4.1	4.5	5.4
40015	4/s Box	ea	.75	.90	1.00	.1	.2	.3
40025	4/0 Box ea	ea	.75	.90	1.00	.1	.2	.3
50030	14 THHN	100/ft	20.00	22.00	30.00	2.8	3.2	4.0
50035	12 THHN	100/ft	25.00	28.00	35.00	2.9	3.3	4.1

Display	Next	Previous	Enter	Change	Remove	Help	Quit

PRICING DATABASE FOR PRICING
AND EXTENSIONS OF THE ESTIMATE

Estimate Number: _____   Job Number: _____   System: _____   System Name: _____

Branch: _____

Section Number: _____   Estimated by: _____   Date: _____

Job Name: _____   Column _____

Code Number	Quantity	Unit	Factor (%)

Display	Next	Previous	Enter	Change	Remove	Help	Quit

Code Number	Description	Unit	Material Price			Labor Units		
			A	B	C	A	B	C

Display	Next	Previous	Enter	Change	Remove	Help	Quit

NAME	**ESTIMATE SUMMARY (COMPUTERIZED)**
NUMBER	**D-2**
ORIGINATOR	**Automatic Computer Output**
COPIES	**3**
DISTRIBUTION	**Chief Estimator, Estimator, General Manager**
SIZE	**8½ × 11 in.**

PURPOSE   The Estimate Summary is generated automatically by the computer using previously entered information. The report is then used by the Chief Estimator and General Manager to determine the bid price.

ESTIMATE SUMMARY

Estimate Number: _____

Job Name: _____ Date:____ / ____ / ____

Job Location: _____

Bid Depository Number: _____ Closing Time: _____ Filing Time: _____

Estimated by: _____ Checked by: _____ Approved by: _____

Sheet Number	Section of Job	Standard Materials	Special Materials	Labor (man-hours)
_____	_____	_____	_____	_____
_____	_____	_____	_____	_____
_____	_____	_____	_____	_____
	_____	_____	_____	_____
Subtotals:	_____

Direct Job Expenses	Job Factor (%)	Subtotal Standard Materials

Bid Bond: _____   Type Building: _____   Miscellaneous Material & Labor: _____

Insurance: _____   Work Conditions:_____   Labor Job Factor, b: _____

Inspection Fee & License: _____   General Contracting: _____   NPL, c: _____

General Supervision: _____   Electrical Contracting: _____   Labor Productivity Factor: _____

Engineering & Drawings:_____   TTL JOB FACT,% _____   Total Material & Labor: _____

Storage & Job Office _____   % NPL (man-hours): _____   Average Labor Rate per Hour

$ _____

Timekeeper: _____   Foreman Supervision: _____   Total $ Value Labor: _____

Freight & Expenses: _____   Study Time: _____   Total Value Material: _____

Truck Expenses: _____   Material Handle: _____   Total Value Direct Job

Expenses, d: _____

Telephone:_____   Order Materials:_____   TOTAL PRIME

COST: _____

NAME	**WEEKLY PAYROLL REPORT**				
NUMBER	**D-3**				
ORIGINATOR	**Automatic computer-generated report**				
COPIES	**1 for each job in progress**				
	1 complete copy for Cost Accounting Department				
DISTRIBUTION	**As noted above**				
SIZE	**8½ × 11 in.**				

PURPOSE By utilizing the computer-generated Weekly Payroll Report, top management and job management are able to tightly control job labor and to identify problem areas almost immediately. The payroll information may be furnished for each job on this report. All the other required current information is obtained automatically by the computer from the current payroll and by accessing job cost records for previous payroll periods by job number to obtain the accumulated totals.

WEEKLY PAYROLL REPORT
Week Ending ____ / ____ / ____

Job Number: _P103091_   Job System Number: _____

Labor Code Number	Total Labor Hours to Date	Total Labor Dollars to Date	Total Estimated Labor Hours	Total Estimated Labor Dollars	To date Over-/Under- estimated	
					Hours	Dollars
2105	40.5	$ 648.00	210.0	$3,360.00	−169.5	−2712.00
2110	200.0	3,200.00	150.0	2,400.00	50.0	800.00
2530	150.0	2,400.00	150.0	2,400.00	0	0
3410	350.0	5,600.00	400.0	6,400.00	−50.0	−800.00

WEEKLY PAYROLL REPORT
Week Ending ____ / ____ / ____

Job Number: _____   Job System Number: _____

Labor Code Number	Total Labor Hours to Date	Total Labor Dollars to Date	Total Estimated Labor Hours	Total Estimated Labor Dollars	To date Over-/Under-estimated	
					Hours	Dollars

NAME **PROGRESS ANALYSIS**

NUMBER **D-4**

ORIGINATOR **Automatic computer-generated report**

COPIES **2**

DISTRIBUTION **Chief Accountant, General Manager**

SIZE **17 × 11 in.**

PURPOSE Each month a comprehensive schedule of work in progress, referred to as the Progress Analysis (Form D-4), is furnished to management.

Such computer-generated reports give management a level of control never before possible. The Progress Analysis report provides management with an accurate picture of the condition of all work in progress. It also enables managers to project future cash flow and determine if the work is proceeding on schedule as projected in the budget. This allows management and front-line supervision to take the steps necessary to correct cost overruns before they become a serious problem.

PROGRESS ANALYSIS

Month: _____

	Man-hours	Labor	Payroll Tax	Material	Direct Job Expense	Tools	Total	Contract Amount	Total cost/ Est Cost = % of Comp.	Project Manager's Estimated % of Completion	Previous Billing	Current % of Complete Billing	Actual Current Billing	Total Billing To Date	Billing Over (Under) Cost & Profit
Job no. 10390															
Estimated	2,000	25,000.00	9,000.00	36,000.00	4,000.00	1,000.00	69,000.00	100,000.00							
Used to date	500	7,000.00	2,240.00	10,000.00	1,500.00	200.00	18,700.00		27%	30%	10,000.00	27,000.00	20,000.00	30,000.00	3,300.00
% used to date	25%	25%	25%	28%	37.5%	20%	27.1%					27%		30%	
Over (under) estimated	(1,500)	(21,000.00)	6,760.00	(26,000.00)	(2,500.00)	(800.00)	(50,300)								
Job no. 23456															
Estimated	5,000	70,000.00	22,400.00	95,900.00	13,000.00	10,300.00	211,600.00	259,300.00							
Used to date	2,500	36,250.00	11,600.00	43,155.00	6,500.00	4,000.00	101,505.00		48%	45%	35,295.00	124,464.00	90,000.00	125,295.00	4,431.00
% used to date	50.0%	51.8%	51.8%	45.0%	50.0%	38.8%	47.9%					48%			
Over (under) estimated	(2,500)	(33,750.00)	(10,800.00)	(52,745.00)	(6,500.00)	(6,300.00)	(110,095.00)								
Job no. _____															
Estimated	—	—	—	—	—	—	—								
Used to date	—	—	—	—	—	—	—								
% used to date	—	—	—	—	—	—	—								
Over (under) estimated	—	—	—	—	—	—	—								
Totals:															
Estimated	7,000	95,000.00	31,400.00	131,900.00	17,000.00	11,300.00	280,600.00	359,300.00							
Used to date	3,000	43,250.00	13,840.00	53,155.00	8,000.00	4,200.00	120,205.00		42.8%		45,295.00	151,464.00	110,000.00	155,295.00	7,731.00
% used to date	42.9%	44.1%	44.1%	40.3%	47.1%	37.2%	42.0%					42.2%		44.2%	
Over (under) estimated	(4,000)	(54,750.00)	17,560.00	(78,745.00)	9,000.00	(7,100.00)	(160,395.00)								
Column number	1	2	3	4	5	6	7	8	9	10	11	12	13	14	15

ESTIMATE SUMMARY

Temporary Power: _____	Testing: _____	OVERHEAD (based on cost), %: _____
Tools & Equipment: _____	Lost Time: _____	OH-Std. Mat.-Lab.-Job. Exp., %: _____
Travel Expenses Days @: _____	TOTAL NPL, c: _____	OH-Special Material & Equipment %: _____
Travel Expenses, mi @: _____	LABOR, % ADDER: _____	TOTAL GROSS COST: _____
Room & Board, days @: _____	FICA: _____	PROFIT, %: _____
Labor Adder, %: _____	SUTA & FUTA: _____	SUBTOTAL: _____
_____	Worker Comp: _____	Sales Tax: _____
_____	PL & PD: _____	Excise Tax: _____
_____	Association Dues: _____	Payment & Performance Bond: _____
_____	NEBF: _____	Interest on financing: _____
_____	Local Health: _____	TOTAL ESTIMATED PRICE: _____
TOTAL JOB EXPENSES, c: _____	TOTAL ADDER, %: _____	AMOUNT OF BID: _____

NPL = nonproductive labor
TTL = total

PROGRESS ANALYSIS

Month: _____

	Man-hours	Labor	Payroll Tax	Material	Direct Job Expense	Tools	Total	Contract Amount	Total cost/ Est Cost = % of Comp.	Project Manager's Estimated % of Completion	Previous Billing	Current % of Complete Billing	Actual Current Billing	Total Billing To Date	Billing Over (Under) Cost & Profit
Job no. 10390															
Estimated															
Used to date															
% used to date															
Over (under) estimated															
Job no. 23456															
Estimated															
Used to date															
% used to date															
Over (under) estimated															
Job no. ____															
Estimated															
Used to date															
% used to date															
Over (under) estimated															
Totals:															
Estimated															
Used to date															
% used to date															
Over (under) estimated															
Column number	1	2	3	4	5	6	7	8	9	10	11	12	13	14	15

ESTIMATE SUMMARY

Temporary Power: _____

Tools & Equipment: _____

Travel Expenses Days @: _____

Travel Expenses, mi @: _____

Room & Board, days @: _____

Labor Adder, %: _____

_____

PL & PD: _____

_____

_____

NEBF: _____

Local Health: _____

TOTAL JOB EXPENSES, d: _____

Testing: _____

Lost Time: _____

TOTAL NPL, c: _____

LABOR, % ADDER: _____

FICA: _____

SUTA & FUTA: _____

Worker Comp: _____

Association Dues: _____

TOTAL ADDER, %: _____

OVERHEAD (based on cost), %: _____

OH-Std. Mat.-Lab.-Job. Exp., %: _____

OH-Special Material & Equipment %: _____

TOTAL GROSS COST: _____

PROFIT, %: _____

SUBTOTAL: _____

Sales Tax: _____

Excise Tax: _____

Payment & Performance Bond: _____

Interest on financing: _____

TOTAL ESTIMATED PRICE: _____

AMOUNT OF BID: _____

NPL = nonproductive labor
TTL = total

CHAPTER FIVE
MAINTENANCE

FORM NO.	FORM NAME	PAGE
MF-1	Tool Register	106
MF-2	Tool Master (Computer Screen)	108
MF-3	Tool Record	110
MF-4	Equipment Maintenance Record	112

NAME	**TOOL REGISTER**
NUMBER	MF-1
ORIGINATOR	Tool Manager
COPIES	2
DISTRIBUTION	Tool Department, Accounting Department
SIZE	11 × 8½ in.

PURPOSE
The ability of electrical contractors to supply the proper tool when it is needed is of equal importance to having the correct material on the job when required. The complexity of present-day electrical installations is such that many tools must be available to the electrical contractor.

As with any investment, the successful company keeps a complete record of the tool inventory. In the case of tools, this is accomplished by creating a Tool Register, which lists all the tools owned by the contractor. Each tool is assigned a number, which identifies it by class and as an individual tool within the class. This may be a number system like the one illustrated on Form MF-1. The listing continues until all tools have been assigned numbers and entered in the computer. A Tool Register, listing all the tools, is then printed and becomes the tool inventory.

TOOL REGISTER

Class	Company Tool Number	Description	Manufacturing Serial Number	Model Number	Quantity	Purchase Price, $	Purchase Date	Replacement Cost, $	Rental Rates, $		
									Daily	Weekly	Monthly
1	1000-23	¾" EMT Bender	12.90099	306	6	250.00	10/09/90	260.00	2.00	6.00	18.00
1	1001-10	1" EMT Bender	23456	602	2	475.00	9/10/89	575.00	5.00	15.00	40.00
2	1020-05	2" GRC Bender	526789	1550	1	2,500.00	1/20/92	3,000.00	20.00	50.00	150.00
6	3550-2	Trencher	98765	32	1	25,000.00	8/10/89	30,000.00	100.00	300.00	900.00

TOOL REGISTER

Class	Company Tool Number	Description	Manufacturing Serial Number	Model Number	Quantity	Purchase Price, $	Purchase Date	Replacement Cost, $	Rental Rates, $		
									Daily	Weekly	Monthly

NAME	**TOOL MASTER (COMPUTER SCREEN)**
NUMBER	**MF-2**
ORIGINATOR	**Tool Manager**
COPIES	**1**
DISTRIBUTION	**Tool Department**
SIZE	**8½ × 5½ in. (Copy form at 100% of original; cut it out along the box rules.)**
PURPOSE	The Tool Master is the screen for entering tool information into the computer, which is then used for the Tool Register.

TOOL MASTER

1. Group _____   2. Suffix _____   3. Company tool no. _____

4. Description _____   5. Serial _____

6. Model no._____   7. Quantity _____

8. Status (new/used) _____   9. Status (date) _____

10. Purchase (date) _____   Purchase price _____

11. Replacement cost _____

12.   <u>Rental rates</u>

    Daily _____   Weekly _____   Monthly _____

Display   **Next**   **Previous**   **Enter**   **Change**   **Remove**   **Help**   **Quit**

NAME	**TOOL RECORD**
NUMBER	**MF-3**
ORIGINATOR	**Tool Department**
COPIES	**1**
DISTRIBUTION	**Tool Department**
SIZE	**8½ × 5½ in.(Copy form at 100% of original;cut it out along box rules.)**

PURPOSE	All the company's major tools require maintenance to keep them in usable condition. The manufacturer generally specifies the type of maintenance required, but it is necessary for the user of the tools to keep a record of maintenance as it is performed.

The Tool Record card is also a record of the date the tool was purchased and the name of the supplier. This information is important in case the contractor needs to take advantage of a warranty.

The record of repairs and maintenance required on the tool can also be an indication of the cost expended for repairs and maintenance. Company policy should state when it is time to replace the tool because of excessive maintenance charges.

TOOL RECORD

Name of Tool: _____ Date Purchased: _____ Vendor: _____

_____

_____

Date	Maintenance Performed	Cost

NAME	**EQUIPMENT MAINTENANCE RECORD**
NUMBER	**MF-4**
ORIGINATOR	**Tool and Equipment Mechanic**
COPIES	**1**
DISTRIBUTION	**Equipment Department**
SIZE	**8½ × 11 in.**

PURPOSE	Probably the largest investment by the contractor is in automobiles, trucks, and major construction equipment. To protect this investment, maintenance must be performed on the proper schedule as designated by the manufacturer. To do this, the Equipment Maintenance Record is necessary.

EQUIPMENT MAINTENANCE RECORD

Name of Equipment: _____

Type of Equipment: _____

Date of Purchase: _____ Purchased from: _____

Lubrication Interval: _____

Maintenance Required: _____ Interval: _____

Maintenance Required: _____ Interval: _____

Maintenance Required: _____ Interval: _____

Date	Miles or Hours	Maintenance Performed	Parts	Cost

CHAPTER SIX
ESTIMATING

FORM NO.	FORM NAME	PAGE
E-1	Job Registration	118
E-2	Specification Information Sheet	120
E-3	Service, Metering, and Grounding Takeoff Sheet	122
E-4	Lighting Fixture Takeoff Schedule	124
E-5	Distribution Equipment Takeoff Schedule	126
E-6	Raceway Equipment Takeoff Schedule	128
E-7	Feeder and Busway Takeoff Schedule	130
E-8	Branch Circuit Takeoff Sheet	134
E-9A	Branch Circuit Takeoff—Shortcut Method	136
E-9B	Branch Circuit Evaluator	138
E-10	Outlet Detail and Takeoff Sheet	140
E-11	Power System Takeoff Schedule	142
E-12	Telephone System Takeoff Schedule	144
E-13	Special Systems Takeoff Schedule	146
E-14	Embedded Tabulation Sheet	148
E-15	Feeder Conduit Tabulating Sheet	150
E-16	Branch Circuit Tabulating Sheet	152
E-17	Conductor Tabulating Sheet	154
E-18	Finishing Tabulating Sheet	156
E-19	Pricing Sheet	158
E-20	Recapitulation Summary Sheet	162
E-21	Bid Summary Sheet	166
E-22	Job Factor Evaluation Sheet	170
E-23	Manpower Chart	172
E-24	Errata Sheet	176

Estimating forms must be interrelated and provide for every step in the estimating process. They must also be structured so that labor units can be assigned properly and factored as required. In addition, they must contain the background information for the job cost accounting involved in job management.

The forms should be so comprehensive that everything involved in the estimate may be included on them, yet so flexible that the estimator can easily make provisions for variations from the printed form.

The procedure followed by the estimator in takeoff work is largely dictated by the layout of the work on the plans, where the design is generally by systems, as follows:

1. Service, metering, and grounding
2. Lighting fixture schedule
3. Equipment schedule
4. Feeder schedule
5. Branch circuit work
6. Power system
7. Telephone system
8. Special systems

Much of the wiring material involved in each of these systems is the same. However, it simplifies listing and pricing to collect these items on a tabulating form so that one item can be priced and labored instead of several entries of the same item. Therefore the tabulating process is important in order to improve accuracy and to simplify pricing.

The tabulating process also makes it possible to accomplish a transition from the order of takeoff by systems to an arrangement of the estimate for job cost accounting through a system which provides a workable arrangement for field reporting according to work schedules, as follows:

1. Embedded work
2. Feeder system
3. Equipment schedule
4. Branch circuit work
5. Conductor schedule
6. Lighting fixture schedule
7. Finishing work

These categories of work represent logical subdivisions of field work to which the wiremen can easily relate in reporting times for their daily work activities. The categories are essential if accurate comparisons are to be made between the estimate and actual field costs. In addition, such a breakdown is useful (and often mandatory) for monthly billings to the owner for reimbursement for completed work.

The design of estimating forms must conform with the concepts in estimating. In the design of the forms used in this book, the basic concept is that the labor units used are standard ones that apply to all forms of estimating, but that must be factored for deviations from standard conditions, which are the following:

✓ A journeyman of average ability using normal tools and procedures will make the installation.
✓ The material will be installed in a localized area of approximately 40,000 sq ft.
✓ The height of installation will not exceed 14 ft above the floor.
✓ Access to the work will be normal for new construction, and it will not be necessary to move obstacles on the floor.
✓ The work will be installed at ground level or similar access.
✓ The work will be installed by one worker as required for one-worker jobs, or two workers as required.
✓ The rate of progress of construction will be normal.
✓ Temperature conditions will be normal and special clothing will not be required.
✓ The concentration of material in a localized area will be normal.

When construction conditions differ from these standards, the following deviations might be made to the standard labor units:

✓ Substandard labor affecting productivity
✓ Labor agreements that require two workers working together for all operations
✓ Work installed more than 14 ft from the floor
✓ Length of wire pulls and of wire pulls in parallel
✓ Conduit runs in parallel
✓ Preassembly and prefabrication
✓ Large concentration of items in a localized area
✓ "Checkerboarding" of the job by the general contractor, or the rate of progress on the job that is not normal, or the job is rushed
✓ Work done over obstacles on the floor
✓ Severe weather conditions, hot or cold
✓ Material furnished by others than the contractor
✓ Conditions that require the workers to wear special clothing

When the estimator identifies the conditions of installation as deviations from standards, it must be easy to transcribe them to the forms for easy recognition for the Pricing Sheet.

NAME	**JOB REGISTRATION**
NUMBER	**E-1**
ORIGINATOR	**Sales Department**
COPIES	**3**
DISTRIBUTION	**Sales, Estimating, General Manager**
SIZE	**8½ × 11 in.**

PURPOSE · This form provides a record of all important information pertaining to the estimate. It:

1. Provides a job name and estimate number that will be used to identify the job throughout its life.
2. Identifies the designers and engineers of the plans, and the name of the contact person.
3. Identifies the source of supply of the plans and the amount of time allowed for usage.
4. Provides written communications to the estimator, salesman, and General Manager.
5. Acts as a permanent record of bids on the job.
6. Serves as a permanent file record of the estimate.

The information provided on the Job Registration form enables the contractor to follow up all matters relating to the bidding processes. The timely receipt and return to the source of supply of blueprints and specifications are important to the supplier, and the disbursement of interoffice information is important to the electrical contractor management.

JOB REGISTRATION

Estimate No.: *142*
Estimator: *R.C.P.*
Job Name: *Littleton Arts Center*
Date: *11-6-XX*
Job Address: *7522 So. Cherokee St.*
City: *Littleton*
Architect: *J. W. Small & Assoc. J. Small*
Contact:
Engineering Firm: *—*
Contact:
Electrical Engineer: *O. K. Briggs*
Contact: *A. Weeks*
Bid Closing Date: *12-1-XX* Time: *3 P.M.* Depository Closing Date: *11-30* Time: *3 P.M.*
Duration of Job: *12 Mo.*
Estimated Man-Hours: *16,000*
Plans Obtained from: *M. K. Ely*
Return Date: *11-20-XX*
Plans Picked up by: *J. Vickers*
Returned by: *J.H.V.*
Deposit Required: *No* Deposit Returned Date: *—*
Returned to: *—*
Special Instructions to Estimator: *Do not include trenching or conc. Provide alt. price on recessed fix.*

Information on Results of Electrical Bidding:

Electrical Contractors	Base Price	Alt.#1	Alt.#2	Alt.#3	Alt.#4
Sturgeon	$390,000.-	+68,000.-	—	—	—
Kennedy	411,000.-	+68,200.-			
Collier	416,000.-	+68,000.-			
Reliable	399,000.-	+68,500.-			

Electrical Awarded to: *Sturgeon @ $452,000.-*

Information on Results of General Contractor Bidding:

General Contractors	Base Price	Alt.#1	Alt.#2	Alt.#3	Alt.#4
M. K. Ely	$2,598,000.-	+64,000.-			
E. B. Jones	2,675,000.-	+66,000.-			
K. C. Const.	3,100,000.-	+66,500.-			
Rocky Mtn.	2,910,000.-	+68,500.-			

General Contract Awarded to: *EBJ*
Notes on Bidding: *Daybrite recessed fixtures approved*
Total Man-Hours Electrical: *17,060*
Maximum Number of Electricians: *7*

JOB REGISTRATION

Estimate No.: _____

Estimator: _____

Job Name: _____

Date: _____

Job Address: _____

City: _____

Architect: _____

Contact: _____

Engineering Firm: _____

Contact: _____

Electrical Engineer: _____

Contact: _____

Bid Closing Date: _____ Time: _____ Depository Closing Date: _____ Time: _____

Duration of Job: _____

Estimated Man-Hours: _____

Plans Obtained from: _____

Return Date: _____

Plans Picked up by: _____

Returned by: _____

Deposit Required: _____ Deposit Returned Date: _____

Returned to: _____

Special Instructions to Estimator: _____

_____

Information on Results of Electrical Bidding:

Electrical Contractors	Base Price	Alt. #1	Alt. #2	Alt. #3	Alt. #4

Electrical Awarded to: _____

Information on Results of General Contractor Bidding:

General Contractors	Base Price	Alt. #1	Alt. #2	Alt. #3	Alt. #4

General Contract Awarded to: _____

Notes on Bidding: _____

Total Man-Hours Electrical: _____ Maximum Number of Electricians: _____

NAME	# SPECIFICATION INFORMATION SHEET
NUMBER	E-2
ORIGINATOR	Estimator
COPIES	2
DISTRIBUTION	Estimator, General Manager
SIZE	8½ × 11 in.
PURPOSE	This form provides a document on which to record the following:

1. General information relative to the job architect contact, bid time and date, bid conditions, bonding and insurance requirements, material substitution allowances, and bidding statistics.

2. Utility service information relative to the point of supply, type of service, service voltage and approximate load requirements.

3. Building construction features relative to the type of floors, walls, ceilings, columns, and inside partitions.

4. Requirements of the wiring system with regard to the type of panelboards, types of wire, switches, the receptacles, plates, and lighting fixtures.

5. Features of the specifications relative to the work of other crafts with regard to patching, cleanup, painting, trenching, concrete, boiler wiring, boiler controls, and temporary wiring to be furnished by the electrical contractor.

Usually the subcontractors are allowed a limited amount of time to use the blueprints and specifications. So this form is used to supply this type of information to the estimator.

SPECIFICATION INFORMATION SHEET

Estimate No.: *148*
Estimator: *R.C.P.*
Date: *11-6-XX*

Job Name: *Littleton Arts Center*

GENERAL

Job Location: *7522 So. Cherokee St.*
Owner: *City of Littleton*
Architect: *Small*   Electrical Engineer: *Briggs*
Bid Date: *12-1-XX*   Completion Date: *12-1-XX*   Retainage: *10%*
Liquidated Damages: *Yes*   Bid Bond: *—*   Special Insurance: *—*
Prime Bid: *—*   Subcontractor Bid: *To G.C.*   Lump Sum: *Yes*
Alternates: *1—Recess Fix.*   Addenda: *—*
Substitution Clause: *Equal as Arch. App.*   Materials Furnished by Others: *Trench & Conc.*
Number of Building: *1*   No./Floors: *1*
Total Floor Area Building: *60,000*   Building: Building:

UTILITY SERVICE

Overhead: *Inside Vault*   Underground: *—*
No./Transformers: *3*   Size in KW: *100*   Location: *Vault*
Location of Meters: *Sw. Rd.*   Vault Work by: *P.S. Co.*
Service Voltage: *120/208*   Phase: *3*   Amps: *830*
Total Lighting Load (KW): *90*   Total Power Load (HP): *200*   Power KW: *150*

TYPE OF CONSTRUCTION

Type of Building: *Comm'l.*   Distance Between Floors: *17 Feet*
Floor Construction: *Slab-2 Fl.*   Ceiling Construction: *7 Bar Susp. (8'-6")*
Outside Wall Construction: *Alum. Skin*   Inside Partition Construction: *Mtl. Stud*
Basement Construction: *No Bam't.*   Slab on Ground: *Yes*
Type of Columns: *Steel with masonry.*

ELECTRIC WIRING

Type of Switchboard: *Molded Case C.B.*
Type of Lighting Panelboards: *Quicklag C.B.*
Type of Power Panelboards: *Frame C.B.*
Type of Wire Branch Circuit: *THW*   Primary: *—*   Feeder: *THW*
Type of Motor Starters: *Mag.*   Type of Control: *Var.*
Type of Wall Switches: *Spec.*   Type of Receptacles: *Spec.*   Type of Plates: *Mtl.*
Color of Wall Switches: *Ivory*   Color of Receptacles: *Ivory*   Color of Plates: *Ivory*
Fixtures Supplied by: *Elec.*   Lamps Supplied by: *Elec.*

MISCELLANEOUS

Patching by: *G.C.*   Cleanup by: *G.C.*   Painting by: *G.C.*
Trenching by: *G.C.*   Concrete Furnished by: *G.C.*
Temporary Wiring by: *Elec. for G.C.*
Boiler Controls by: *Mech.*   Boiler Wiring by: *Elec.*
Motor Starters by: *Elec.*   Control by: *Mech.*
Drawings Checked by: *R.C.P.*   Specifications Checked by: *R.C.P.*
Job Site Checked by: *R.C.P.*   Date of Visit: *11-7-XX*

SPECIFICATION INFORMATION SHEET

Estimate No.: _____

Estimator: _____

Date: _____

Job Name: _____

GENERAL

Job Location: _____

Owner: _____

Architect: _____ Electrical Engineer: _____

Bid Date: _____ Completion Date: _____ Retainage: _____

Liquidated Damages: _____ Bid Bond: _____ Special Insurance: _____

Prime Bid: _____ Subcontractor Bid: _____ Lump Sum: _____

Alternates: _____ Addenda: _____

Substitution Clause: _____ Materials Furnished by Others: _____

Number of Building: _____ No./Floors: _____

Total Floor Area   Building: _____ Building: _____ Building: _____

UTILITY SERVICE

Overhead: _____ Underground: _____

No./Transformers: _____ Size in KW: _____ Location: _____

Location of Meters: _____ Vault Work by: _____

Service Voltage: _____ Phase: _____ Amps: _____

Total Lighting Load (KW): _____ Total Power Load (HP): _____ Power KW: _____

TYPE OF CONSTRUCTION

Type of Building: _____ Distance Between Floors: _____

Floor Construction: _____ Ceiling Construction: _____

Outside Wall Construction: _____ Inside Partition Construction: _____

Basement Construction: _____ Slab on Ground: _____

Type of Columns: _____

ELECTRIC WIRING

Type of Switchboard: _____

Type of Lighting Panelboards: _____

Type of Power Panelboards: _____

Type of Wire   Branch Circuit: _____ Primary: _____ Feeder: _____

Type of Motor Starters: _____ Type of Control: _____

Type of Wall Switches: _____ Type of Receptacles: _____ Type of Plates: _____

Color of Wall Switches: _____ Color of Receptacles: _____ Color of Plates: _____

Fixtures Supplied by: _____ Lamps Supplied by: _____

MISCELLANEOUS

Patching by: _____ Cleanup by: _____ Painting by: _____

Trenching by: _____ Concrete Furnished by: _____

Temporary Wiring by: _____

Boiler Controls by: _____ Boiler Wiring by: _____

Motor Starters by: _____ Control by: _____

Drawings Checked by: _____ Specifications Checked by: _____

Job Site Checked by: _____ Date of Visit: _____

NAME	SERVICE, METERING, AND GROUNDING TAKE-OFF SHEET
NUMBER	E-3
ORIGINATOR	Estimator
COPIES	2
DISTRIBUTION	Job Processor, Estimator
SIZE	8½ × 11 in.
PURPOSE	This document contains the following takeoff information on all features of the main service up to but not including the switchboard:

1. Primary source of supply with regard to crib or vault detail if part of the electrical contractor's work.
2. Service conductor data from the transformers secondary to the switchboard.
3. Service raceway information from the transformers secondary to the switchboard.
4. Meter housing information, together with current transformer housing and meter wire leads.
5. Grounding information relative to type of electrode, clamp, cable, and lugs.
6. Items of equipment such as capacitors.

This form becomes one of the job estimating documents, and the takeoff information is posted on the Pricing Sheet (Form E-19) under the work schedule caption for Equipment. The wire detail is posted on the same form under the work schedule caption for Conductor. This information becomes part of the Bill of Material separate caption for Equipment 2 and a separate caption of Conductor.

SERVICE, METERING, AND GROUNDING TAKEOFF SHEET

Sheet Number: _1_ of _1_
Estimator: _R.C.P._
Job Name: _Littleton Arts Center_   Date: _11-6-92_

Material Item	Description	Quantity	Total
SOURCE			
Transformer, Size & Type	By utility	—	
Voltage, Primary			
Voltage, Secondary			
Arresters			
Cutouts, or Switches			
Insulators			
Hardware			
Wire, Type & Size	↓		
SERVICE			
Conductor, Type & Size	500 MCM–T.H.W.	68	278
Conductors, Number of	4–2 runs in par'l.		
Entrance Fitting, Type & Size	"A" Conduit	2	
Length of Wire Pull		26	
Terminals, Wire Size	500 MCM Lugs	8	8
	Burndy Conn	8	8
RACEWAY			
Raceway, Type & Size	3½" G.R.S.		
Length of Run	2 Runs	26	58
Type of Construction	Exp. on Masonry		
Number Runs in Parallel	2		
Elbows	3½"–90°	2	4
Fittings	"A" Conduit & 44. cov.	2	2
Offsets or Bends	No	—	
Terminals, Conduit	4 L.N., 2 Bush.	2	4
METERING			
Housing, Type	1–Type B	1	1
Cabinet, Current Transformer	24" × 36" C.T. Cab.	1	1
Meter Leads Conduit Size	1" E.M.T.	20	20
Wires, Size	#10 THW–7 Cond.	10	70.
GROUNDING			
Clamp, Type & Size	Clamp–App.	1	1
Cable, Type & Size	#0 T.H.W.	38	38
Electrode, Type	Water Pipe–4" O.D.	1	1
Terminals, Wire	Size 0–Sols. Lug	1	1
Conduit	¾" E.M.T.	30	30
Bushing	3/4 Grnd. Bush	1	1
MISCELLANEOUS			
Bushing	3/4 Grnd. Bush	1	1

SERVICE, METERING, AND GROUNDING TAKEOFF SHEET

Sheet Number: _____ of _____

Estimator: _____

Job Name: _____

Date: _____

Material Item	Description	Quantity	Total
SOURCE			
Transformer, Size & Type			
Voltage, Primary			
Voltage, Secondary			
Arresters			
Cutouts, or Switches			
Insulators			
Hardware			
Wire, Type & Size			
SERVICE			
Conductor, Type & Size			
Conductors, Number of			
Entrance Fitting, Type & Size			
Length of Wire Pull			
Terminals, Wire Size			
RACEWAY			
Raceway, Type & Size			
Length of Run			
Type of Construction			
Number Runs in Parallel			
Elbows			
Fittings			
Offsets or Bends			
Terminals, Conduit			
METERING			
Housing, Type			
Cabinet, Current Transformer			
Meter Leads Conduit Size			
Wires, Size			
GROUNDING			
Clamp, Type & Size			
Cable, Type & Size			
Electrode, Type			
Terminals, Wire			
MISCELLANEOUS			

NAME	**LIGHTING FIXTURE TAKEOFF SCHEDULE**
NUMBER	**E-4**
ORIGINATOR	**Estimator**
COPIES	**2**
DISTRIBUTION	**Estimator, Processor**
SIZE	**17 × 11 in. (Copy form at 170% of original.)**

PURPOSE   The Lighting Fixture Takeoff Schedule documents all takeoff information for lighting fixtures, lamps, hanging details, and manufacturer, as well as all other information required for the job, as follows:

1. Types of fixtures shown on the plans according to plan designations, together with the specified manufacturer's name and catalog number, as shown on the plans or in the specifications.
2. Provisions are made on the form to show fixture mounting details, hanger information, wattage, and voltage.
3. Lamp requirements for each fixture as to type, color, and wattage.
4. Provisions are made to show information relative to aligners, dimmers, lowering devices, or timers.

During takeoff, the estimator must designate the items, if any, that are installed at a height above 14 ft (as provided for in the top section of the form) for the purpose of factoring the labor required to install them. One way to do this is to place a dot in the space for the quantity of fixtures that are subject to factoring for a height up to 25 ft. Mark an X in the space with the quantity to be installed up to 35 ft. These designations are used for factoring when the takeoff information is placed on the Pricing Sheet. This form may be used in the field during construction. The information on it is posted directly to the Pricing Sheet (Form E-19).

LIGHTING FIXTURE TAKEOFF SCHEDULE

Job Name:

Sheet Number: _____ of _____
Floor Number: _____ Building: _____
Estimator: _____
Date:

Signals Used on Forms for Rigging Deviations from Standards

Height of Installation		Conduit Parallel Runs		Wire Pulls in Parallel		Miscellaneous Signals
Height	Designation	Number	Designation	Number	Designation	
15–25 ft	o	2	2r	2	2p	Density of items—d
26–35 ft	x	3	3r	3	3p	
36–50 ft	✓	4	4r	4	4p	
		5	5r	5	5p	Number of Men—n

Note: Designation should be made in upper or lower left-hand corner of quantity column square.

TEM Number	Plan Designation	Quantity	Description	Type Fixture	Spec. Mfg.	Mfg. Cat. No.	Fix. Mount.	Watts	Volts	Hanger	Hgr. Const.	Lamps					Aligned	Dimmer	Lowering Device	Timing Device
												No.	Type	Color	Watts Ea.					

DISTRIBUTION EQUIPMENT TAKEOFF SCHEDULE

NUMBER **E-5**

ORIGINATOR **Estimator**

COPIES **2**

DISTRIBUTION **Estimator, Job Processor**

SIZE **17 × 11 in. (Copy form at 170% of original.)**

PURPOSE This form documents the takeoff information relative to the electrical distribution equipment required on the plans or specifications:

1. Provisions are made to describe the main load centers, including the main breaker, related panel sections, and feeders, together with wire terminals and conduit entrances

2. Panelboards fed by each feeder, showing the number of branches, the size of each breaker, and the number of poles on each breaker

3. Safety switches that are fed from branch circuit panelboards or directly from the main load center

4. Individual enclosed circuit breakers, as they are fed from branch circuit panelboards

5. Contactors or motor starters that are fed from individual safety switches or circuit breakers

6. Other equipment items, such as capacitors

The information on this form should be posted directly to the Pricing Sheet (Form E-19), Equipment Work Schedule. In addition to the information transcribed from Form E-5, there will be other items of equipment from other takeoff sheets, but tabulating is not required.

DISTRIBUTION EQUIPMENT TAKEOFF SCHEDULE

Sheet Number: _____ of _____
Floor Number: _____ Building: _____
Estimator: _____
Date: _____

Job Name: _____

		Loadcenter—Type:								Panelboards—Type:				Mains:	Volts:		Safety Switches—Volts:						Circuit Breakers (encl.)						Contactors (encl.)		
	Volts:	Mains:		C.T.S.				Panel Desig.	Type Branch	Mounting	Single Pole	Two Pole	Three Pole			Quan.	Amps	Type	Poles	Mtg.		Quan.	Amps	Type	Poles	Mtg.		Quan.	Amps	Poles
No.	Type	Poles	Amps	W.Term.	C. Entr.						20A.																				

NAME **RACEWAY EQUIPMENT TAKEOFF SCHEDULE**

NUMBER **E-6**

ORIGINATOR **Estimator**

COPIES **2**

DISTRIBUTION **Estimator, Job Processor**

SIZE **17 × 11 in. (Copy from 170% of original.)**

PURPOSE On this document is recorded the takeoff information relative to the quantity and type of raceway equipment. *Raceway equipment* is defined as any type of equipment other than conduit that provides a raceway for wire or cable:

1. The first section of the form is devoted to Unistrut type channel and fittings. While this type of channel may be used as a raceway, it is generally used as a structural material for supporting conduits, cable tray, or lighting fixtures.

2. The second section is devoted to recording quantities and types of tray or fittings.

3. Third, the form records the quantities and types of underfloor duct and fittings, as follows:

    a. Underfloor duct and fittings

    b. Header duct and fittings

    c. Trench duct and fittings

4. The fourth section is devoted to surface duct or the Wiremold type of raceway and fittings.

5. The fifth section of the form is devoted to the quantities and types of wireway or gutter and fittings.

6. The last section is undesignated so that the estimator can use the space for other items not specifically covered by the form.

The information on this form is posted directly to the Pricing Sheet (Form E-19), Equipment Work Schedule.

RACEWAY EQUIPMENT TAKEOFF SCHEDULE
(USED FOR TAKEOFF OF CHANNEL, TRAY, DUCT, MOLD & WIRE NAV)

Job Name: _____

Sheet Number: _____ of _____

Floor Number: _____ Building: _____

Estimator: _____

Date: _____

Channel (Unistrut)

Quan.	Size	Type	Angle	Bar	Beam Clamp		Rod		Cond. Straps		Spring Nuts	
					Quan.	Size	Quan.	Size	Quan.	Size	Quan.	Size

Cable Support (Tray)

Quan.	Size	Type	Cover	Flat El	Vert. El	Splice Plate	Hanger	Quan.	Type	Comb.	El	Off. El	Vert. El	Cab Conn.	Cond. Adap. Quan.	Size	Serv. Fig. Quan.	Type	Junc. Box Quan.	Type	Support

Underfloor Duct—Header Duct—Trench Duct

Wiremold and Fittings

Quan.	Size	Type	Out Box	Ext. Box	Junc. Box	Fix. Box	Ext. El	Flat El	Int. El	End Cover	Fitting	Recep.	Switch

Wireway (Gutter)

No. Pieces	Length	Quan.	Size	Type	Cover	Conn.	End Clo.	Hanger	Tee	Panel Entrance	Quan.	Volts	Amps	Ends	Hangers

NAME	**FEEDER AND BUSWAY TAKEOFF SCHEDULE**
NUMBER	**E-7**
ORIGINATOR	**Estimator**
COPIES	**2**
DISTRIBUTION	**Estimator, Job Processor**
SIZE	**17 × 11 in. (Copy form at 170% of original.)**
PURPOSE	On this document is recorded the takeoff information relative to the lighting and power feeders, including wire, cable, and busway:

1. The first section of the form is devoted to feeder wire, cable, or busway. The first four columns identify the feeder number shown on the plans, its source, destination, and a description (wire, cable, or busway).

   The next five columns include the size in circular mils (or amperage), the number of conductors or busway bars, the length of the run including makeup, the type of insulation or type of busway, and the number of terminals.

2. The first two columns describe the size and type of raceway.

3. The next five columns describe the length of run posted in a column headed by the type of construction in which it is installed. The length of the raceway is recorded in even feet to compensate for the length of the product being used. During takeoff, it is necessary for the estimator to designate the height of installation, runs in parallel, or runs in trench, for factoring as follows:

Height of installation:	15–25 ft (mark * by footage)
	26–35 ft (mark x by footage)
	36–50 ft (mark with dot)
Runs in parallel:	Mark with r in front of footage, such as "2r180," designating two runs in parallel.
Pulls in parallel:	Designate with a small p.
Conduit in trench:	Show a small t in front of the footage.
Conduit risers:	Designated by R.

4. The last column in this section is used to record the conduit terminals in the run.

5. The last section on the form is used to record the quantity, type, and size of fittings for the raceway and busway.

   Several columns in the last section are headed for specific fittings, but several are left blank for fittings other than those designated.

The information relative to wire and cable is posted on Form E-17.

The information on busway is posted on Form E-19, Equipment Work Schedule.

The raceway information is posted on the Feeder Conduit Tabulating Sheet (Form E-15).

Note: A separate Form E-15 is required for each different type of raceway.

FEEDER AND BUSWAY TAKEOFF SCHEDULE

Sheet Number: _1_ of _1_
Floor Number: _1_ Building: _A_
Estimator: _R.C.P._
Date: _11-7-92_

Job Name: _Littleton Arts Center_

	Signals Used on Forms for Flagging Deviations from Standards						
Height of Installation		Conduit Parallel Runs		Wire Pulls in Parallel		Miscellaneous Signals	
Height	Designation	Number	Designation	Number	Designation		
15–25 ft	•	2	2r	2	2p	Density of Items—d	
26–35 ft	x	3	3r	3	3p		
36–50 ft	o	4	4r	4	4p	Number of Men—n	
		5	5r	5	5p		

Note: Designation should be made in upper or lower left-hand corner of quantity column square.

Feeder No.	From	To	Desc.	Size	No. of Wires (for Conductors)	Length of Pull	Type	Wire Term.	Size	Type	Embedded	Trench	Exposed	Hanger	Riser	Cond. Terminal	Ells	Fitting	Hanger	End Closure	Junc. Boxes	Tees	Cross	C.B. Plug	V.B. Plug
2.	Sw. Bd.	LPA	Wire	250	4	2p 181'	THW	4	2½"	GRS				2p180'		4	2	1-LB	2 Cond. Trap	–	1-30 x 30 x 6	–			–
3.	Sw. Bd.	LPB	Cable	2/0	4	86'	RR	4	2"	GRS	80'					2	2	–		–	–	–			–
4.	Sw. Bd.	LPC	Wire	250	4	136'	THW	4	2½"	GRS	120'					2	2	–		–	–	–			–
5.	Sw. Bd.	PPA	Cable	4/0	3	65'	RR	3	2"	GRS	80'	60'				2	1	1-LL		–	1-16 x 18 x 6	–			–
6.	Sw. Bd.	PPB	Wire	1/0	3	2p 146'	THW	3	1½"	GRS				2p 140'		4	2	–	2 Cond. Trap	–	–	–			–
7.	Sw. Bd.	Air Cond	Wire	4/0	3	80'	THW	3	2"	GRS			80'			2	2	–	–	–	–	–			–
8.	Sw. Bd.	Shop	BUS	1000A	3	400'	Plug-In	2	–	–				400'		1-BUS	4	C.B. Incl.	40	1	–	–	–		10

FEEDER AND BUSWAY TAKEOFF SCHEDULE

Job Name: _____

Sheet Number: _____ of _____
Floor Number: _____ Building: _____
Estimator: _____
Date: _____

Signals Used on Forms for Flagging Deviations from Standards

Height of Installation		Conduit Parallel Runs		Wire Pulls in Parallel		Miscellaneous Signals
Height	Designation	Number	Designation	Number	Designation	
15–25 ft	*	2	2r	2	2p	Density of items—d
26–35 ft	x	3	3r	3	3p	
36–50 ft	o	4	4r	4	4p	
		5	5r	5	5p	Number of Men—n

Note: Designation should be made in upper or lower left-hand corner of quantity column square.

Feeder No.	Feeder Wire, Cable, or Busway							Raceway						Length of Run						Raceway–Busway Fittings								
	From	To	Desc.	Size	No. of Wires (for Conductors)	Length of Pull	Type	Wire Term.	Size	Type	Embedded	Trench	Exposed	Hanger	Riser	Cond. Terminal	Ells	Fitting	Hanger	End Closure	Junc. Boxes	Tees	Cross	C.B. Plug	V.B. Plug		

NAME	**BRANCH CIRCUIT TAKEOFF SHEET**
NUMBER	**E-8**
ORIGINATOR	**Estimator**
COPIES	**2**
DISTRIBUTION	**Estimator, Job Processor**
SIZE	**17 x 11 in. (Copy form at 170% of original.)**
PURPOSE	The Branch Circuit Takeoff Sheet records the quantities and types of raceway and wire or cable used to install the branch circuit wiring for fixture outlets, switches, receptacles, and special purpose outlets:

1. The first section of the form provides for recording quantities of raceway according to sizes ½ in. and ¾ in., as installed in one of three types of construction:

   a. Exposed on surface

   b. Furred ceiling

   c. Frame-metal or wood

   Since the lengths are measured between outlets, the footage is recorded in the appropriate column according to the type of construction in which it is installed.

2. The next section of the form provides for recording the wire contained in the raceway. The lineal lengths are recorded under the proper sizes in a column under the number of conductors. The total footage of wire is not recorded.

3. The third section of the form provides for recording the takeoff of nonmetallic sheathed cable, which may be used for certain types of branch circuit wiring.

4. Footage of the raceway and wire is totaled at the bottom of the sheet.

The total footage of raceway and wire is posted on the Branch Circuit Tabulating Sheet (Form E-16) for tabulating with other materials of a similar nature.

BRANCH CIRCUIT TAKEOFF SHEET

Signals Used on Forms for Flagging Deviations from Standards

Height of Installation		Conduit Parallel Runs		Wire Pulls in Parallel		Miscellaneous Signals
Height	Designation	Number	Designation	Number	Designation	
15–25 ft	o	2	2r	2	2p	Density of items—d
26–35 ft	x	3	3r	3	3p	
36–50 ft	✓	4	4r	4	4p	Number of Men—n
		5	5r	5	5p	

Note: Designation should be made in upper or lower left-hand corner of quantity column square.

Sheet Number: _____ of _____   Building: _____
Floor Number: _____
Estimator: _____
Date: _____

Job Name: _____

		Raceway																			Wire—Size and Number of Conductors																Cable R/Ground Wire			
		½"						½"						¾"						#12					#12					#10				Type	12-2	12-5	6-3	Term
Ct. No.	Type	Expd.	Furred	Frame	Term	Exp.	Furred	Frame	Term	Exp.	Furred	Frame	Fitting	Term.	Type	2	3	4	5	2	3	4	5	2	3	4	5	2	3	4							

NAME	# BRANCH CIRCUIT TAKEOFF—SHORTCUT METHOD
NUMBER	**E-9A**
ORIGINATOR	**Estimator**
COPIES	**I**
DISTRIBUTION	**Estimator**
SIZE	**11 x 8½ in.**

PURPOSE

This form provides a data table with an appropriate graph. From them, you can determine the branch circuit conduit and wire quantities for an area according to the number of outlets involved in the area. The assumption is that the layout configuration of branch circuit outlets follows a uniform pattern, such as it would in any structure with a number of rooms of identical size that are to be used for the same purpose. The form is arranged as follows:

1. The first section is used in connection with the Evaluator Curve (Form E-9B), which enables the estimator to determine the average number of feet of conduit per outlet of branch circuit conduit from:

   a. The area data of the space.

   b. The number of outlets in the space.

2. The next section of the form enables the estimator to determine the footages of ½-in. and ¾-in. conduit, depending on whether the wiring system is 3-wire or 4-wire, and on whether the switching is local or panel. These determinations are the result of evaluating the conduit feet per outlet from the graph on Form E-9B, together with the number of outlets.

3. The third section is used for determining the number of feet of wire in a 3-wire system. Multiply the total feet of conduit by the percentages shown on the form, depending on whether the switching is local or panel.

4. The last section of the form is used to determine the footage of wire in a 4-wire system. Again, multiply the total footage of conduit by the percentages shown on the form, depending on whether the switching is local or panel.

The information determined from this form is posted on Form E-16, Branch Circuit Tabulating Sheet, for tabulating with material of a similar type from other systems.

BRANCH CIRCUIT SYSTEM—SHORTCUT METHOD

Job Name: _____

DATA TABLE FOR CALCULATIONS—SEE GRAPH FORM 9B

Area No.	Ceiling Height			Area Data—L/W Ratio					Area (sq ft)	Quan. Outlets	Sq Ft Outlet	Switching at Panel		Panel Location		Graph Data	
	A-9 ft	B-12 ft	C-15 ft	Length	Width	A-2.5/1	B-4/1	C-6/1				B	C	A	B	Curve	Ft/Outlet

DETERMINATION OF FOOTAGE OF RACEWAY FOR 3-WIRE AND 4-WIRE SYSTEMS

Type of Conduit	Conduit Feet per Outlet	Quantity of Outlets	Total Feet of Conduit	3-Wire System						4-Wire System					
				Local Switching			Panel Switching			Local Switching			Panel Switching		
				Size	Ratio of Sizes	Quantity	Size	Ratio of Sizes	Quantity	Size	Ratio of Sizes	Quantity	Size	Ratio of Sizes	Quantity

DETERMINATION OF FOOTAGES OF WIRE FOR A 3-WIRE SYSTEM—TYPE OF WIRE _____

Total Feet of Conduit	Local Switching					Panel Switching				Multiplier	Quantity of Wire	Plus 5% Makeup	Quantity Lineal Footage of Wire
	2-#12 57%	3-#12 24%	4-#12 5%	5-#12 11%	6-#12 3%	2-#12 65%	3-#12 16%	4-#12 2%	5-#12 17%				

*Place figures in these spaces.

DETERMINATION OF FOOTAGE OF WIRE FOR 4-WIRE SYSTEM—TYPE OF WIRE _____ TOTAL _____

Total Feet of Conduit	Local Switching					Panel Switching				Multiplier	Quantity of Wire	Plus 5% Makeup	Quantity Lineal Footage of Wire
	2-#12 57%	3-#12 25%	4-#12 12%	5-#12 3%	6-#12 3%	2-#12 61%	3-#12 21%	4-#12 16%	5-#12 2%				

*Place figures in these spaces.

NAME	**BRANCH CIRCUIT EVALUATOR**
NUMBER	**E-9B**
ORIGINATOR	Estimator
COPIES	1
DISTRIBUTION	Estimator
SIZE	11 × 8½ in.

PURPOSE   This graph provides a means of determining the conduit feet per outlet based on the square feet of room space per outlet. This graph is used in connection with Form E-9A for determining conduit feet and wire quantities. The conduit feet per outlet (in the first section of Form E-9A, Data Table) provides the basis for the use of the Evaluator.

Of three curves on the evaluator, any one may be used in accordance with the characteristics of the room relative to ceiling height and ratio of length to width, as follows:

1. Curve A is used when there is a maximum ceiling height of 9 ft, and the ratio of room length to width is 2.5 to 1.
2. Curve B applies when there is a maximum ceiling height of 12 ft, and the ratio of length to width is 4 to 1.
3. Curve C applies when the ceiling height is a maximum of 15 ft, and the ratio of length to width is 6 to 1.

The values represented by these curves is based on local switching. When switching is provided at the panelboard, move up one curve.

Likewise, if the panelboard is located in the corner of the room or at the end of a rectangular-shaped space, move up one curve.

Since this is a reference form, no information is recorded on it. And no information is transferred to another form, other than the reference data used on Form E-9A to determine the footage of conduit and wire used in branch circuit installation work.

BRANCH CIRCUIT EVALUATOR

Note: For switching at panel or panel off room center, move up one.

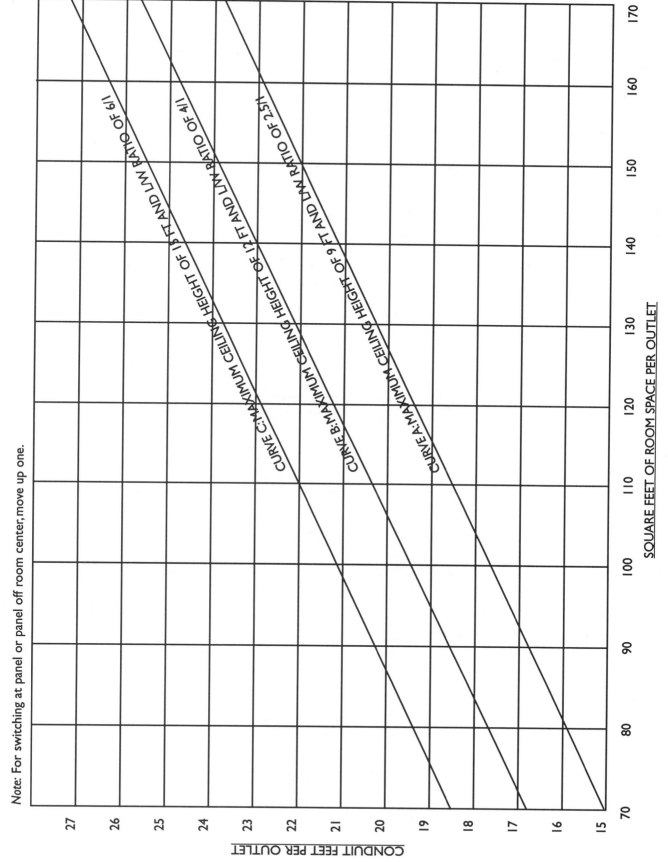

CURVE C: MAXIMUM CEILING HEIGHT OF 15 FT AND L/W RATIO OF 6/1

CURVE C: MAXIMUM CEILING HEIGHT OF 12 FT AND L/W RATIO OF 4/1

CURVE B: MAXIMUM CEILING HEIGHT OF 9 FT AND L/W RATIO OF 2.5/1

CURVE A: MAXIMUM CEILING HEIGHT OF

CONDUIT FEET PER OUTLET

SQUARE FEET OF ROOM SPACE PER OUTLET

NAME	**OUTLET DETAIL AND TAKEOFF SHEET**
NUMBER	**E-10**
ORIGINATOR	**Estimator**
COPIES	**2**
DISTRIBUTION	**Estimator, Job Processor**
SIZE	**17 × 11 in. (Copy form at 170% of original.)**

PURPOSE   This document records the quantities, types, and type of construction in which they are installed, for all types of branch circuit outlets. A list of the materials required to install outlets is therefore obtainable from this form. A different material composition is required to install each type of outlet in each type of construction, with a resultant accurate bill of material.

After counting the number of outlets in a type of construction, indicate the number in the cross-section space for each type of material required.

Provisions are made at the bottom of each column to summarize all the items of material involved in the group. This summary should include two conduit terminals for each type of outlet (or nonmetal sheath cable). In addition, all ceiling and bracket outlets should include 21/2 wire terminals.

There are two sections of the form for material lists which are summarized at the bottom of each column. One is for material that is installed during branch circuit rough-in, and the other is for material used in finishing.

The material list summarized at the bottom of form is divided into two sections, as follows:

1. Material list showing items of material which will be posted on Form E-16, Branch Circuit Tabulating Sheet

2. Material list showing items of material that will be posted on Form E-18, Finishing Tabulating Sheet.

OUTLET DETAIL AND TAKEOFF SHEET

OUTLET DETAIL AND TAKEOFF SHEET

Sheet Number: _____ of _____
Floor Number: _____ Building: _____
Estimator: _____
Date: _____

Job Name: _____

Rough-in Material

Finishing Material

Type	Description	Quantity	4 sq.in. Box-1½	4 in.Oct. Box-1½	4¹¹⁄₁₆ sq.in. Box-2⅛	4 sq.in. Box-2⅛	4 in.Oct. Box-2⅛	Steel Flr Box	Arch. Plas.	Ass'y Bar Form	Tog. Bolt	Bushing GRC ½"	Bushing GRC ¾"	Bushing GRC 1"	Conn Wire	Conn. EMT ½"	Conn. EMT ¾"	Conn. EMT 1"	Cov. 4 sq.in. Blnk	Cov. 4 sq.in. Incl.1G	Cov. 4 sq.in. Incl.2G	Cov. Plas. Dev.	Cov. 3 in. Oct.	Cov. WPF S/W	Hang Box J Bar	Hang Box Adj.	Hang Box Deep	Locknut ½"	Locknut ¾"	Locknut 1"	Plate, Dev. KO Blnk	D.R.	Sw.	Tel.	Recep. Switch Duplex	Sing	I P	3W	Scr 1"-8	Stud Fix	Misc.
Ceiling	Outlet-Fixture	Exposed on Concrete																																							
		Exposed on Furred																																							
		Flush																																							
		Recessed in Furred	//																																						
		Wood frame Flush																																							
	Wall	Masonry flush																																							
		Metal stud flush																																							
		Rein. concrete flush																																							
Receptacle	Duplex	Exposed on concrete	/																																						
		Masonry flush	///																																						
		Metal stud flush																																							
		Wood frame flush																																							
	Single	Deck flush																																							
		Rein. concrete flush																																							
		Wood frame flush																																							
Switch	Tumbler-Wall	Exposed on concrete	/																																						
		Masonry flush	//																																						
		Metal stud flush																																							
		Wood frame flush																																							
		Rein. concrete flush																																							
Junction	Wall	Metal stud																																							
		Reinforced concrete																																							
		Wood frame																																							
Telephone	Floor Wall	Masonry flush																																							
		Wood frame flush																																							
		Deck flush																																							
		Totals																																							

NAME	**POWER SYSTEM TAKEOFF SCHEDULE**
NUMBER	**E-11**
ORIGINATOR	**Estimator**
COPIES	**2**
DISTRIBUTION	**Estimator, Job Processor**
SIZE	**17 × 11 in. (Copy form at 170% of original.)**

PURPOSE

This schedule documents the take-off quantities of equipment, polarized plug caps, polarized receptacles, controls, raceway, and conductor used in the installation of electrical work for motors, power outlets, fans, heaters, and branch circuit work associated with the power system:

1. Each item of equipment is identified by name, source of supply, circuit number, horsepower, phase, volts, and supplementary equipment items.
2. A section of the form is devoted to a description of power outlets as to amperage, voltage, poles, and type of insulation.
3. The next section is devoted to raceway and fittings. The length of run is rounded out to even lengths, to compensate for construction irregularities and trade product lengths. The length of run is posted in the column that represents the type of construction in which it is installed. Provisions are made for boxes for power outlets in the fittings section.
4. The last section is devoted to conductor, and lengths are recorded in lineal feet according to the length run plus makeup.

The information gathered on this form is posted to three different tabulating forms and one Pricing Sheet "work schedule," as follows:

1. Power equipment items are posted to Pricing Sheet (Form E-19), Equipment Work Schedule.
2. Power outlets are posted to Finishing Tab (Form E-18).
3. Branch circuit raceway and fittings are posted to Branch Circuit Tabulating Sheet (Form E-16).
4. Feeder raceway and fittings are posted to Feeder Conduit Tabulating Sheet (Form E-15).

Much of the work recorded on this form is branch circuit work, but it must be recorded on this form. Designations must be made for the height of installation above 14 ft, and these are made in the same manner as all other takeoffs.

POWER SYSTEM TAKEOFF SCHEDULE

Signals Used on Forms for Flagging Deviations from Standards

Height of Installation		Conduit Parallel Runs		Wire Pulls in Parallel		Miscellaneous Signals
Height	Designation	Number	Designation	Number	Designation	
15–25 ft	o	2	2r	2	2p	Density of items—d
26–35 ft	x	3	3r	3	3p	
36–50 ft	✓	4	4r	4	4p	
		5	5r	5	5p	Number of Men—n

Note: Designation should be made in upper or lower left-hand corner of quantity column square.

Sheet Number: _____ of _____

Floor Number: _____

Estimator: _____

Date: _____

Job Name: _____

Identification			Motor Service								Power Outlets					Conductor				Raceway				Length Run						Ells	Fitting	Boxes	Term.
Source Supply	Ct. No.	Location	H.P.	PH.	Volts	Safety Switch	Starter	P.B. Sta.	Flex. Conn.		Amps	Volts	Poles	Fl. or Surf.	Size	Type	No. of Wires	Length of Pull	Term	Size	Type	Emb'd	Exp	Furred	Frame	Hanger							

NAME **TELEPHONE SYSTEM TAKEOFF SCHEDULE**

NUMBER **E-12**

ORIGINATOR **Estimator**

COPIES **2**

DISTRIBUTION **Estimator, Job Processor**

SIZE **17 × 11 in. (Copy form at 170% of original.)**

PURPOSE On this document record the quantities of material relating to the installation of telephone panelboards, raceway, outlets, and all other items involved in the telephone system:

1. The first section provides for listing the telephone panelboards by number, size, and mounting.

2. The next section provides for listing the conduit runs. The length of the run is posted in the proper column in even footage according to the type of construction in which it is installed. Raceway fittings are listed at the end of this section.

3. The third section is devoted to recording the quantity of outlets, including a description of the type and mounting conditions. The estimator uses Form E-10 to determine the detail involved in the mounting material for the outlet box and hanger.

4. The fourth section provides for listing details of the wire required to make the installation. When the electrical contractor does not furnish or install the telephone wire, it is usually necessary for the contractor to install a #8 steel fishwire.

5. The end section provides for listing such things as terminal blocks.

From this form the estimator derives information to list panelboards on the Pricing Sheet (Form E-19), Equipment Work Schedule.

Feeder conduit takeoff is posted on Feeder Tabulating Sheet (Form E-15).

Branch circuit conduit is posted on Branch Circuit Tabulating Sheet (Form E-16).

Outlets should be detailed on Form E-10, and then posted on Branch Circuit Tabulating Sheet (Form E-16). Wire should be posted on Conductor Tabulating Sheet (Form E-17).

Most telephone system hookups are installed by the utility. The contractor installs all the rough-in materials, and the utility pulls the cables and installs the interiors of the panels. The contractor installs a #8 steel fishwire. The type of construction and height of installation must be recognized by the estimator when evaluating the labor required to make the installation.

TELEPHONE SYSTEM TAKEOFF SCHEDULE

Job Name: _____

Sheet Number: ____ of ____
Floor Number: ____ Building: ____
Estimator: ____
Date: ____

Signals Used on Forms for Flagging Deviations from Standards

Height of Installation		Conduit Parallel Runs		Wire Pulls in Parallel		Miscellaneous Signals
Height	Designation	Number	Designation	Number	Designation	
15–25 ft	o	2	2r	2	2p	Density of items—d
26–35 ft	x	3	3r	3	3p	
36–50 ft	✓	4	4r	4	4p	Number of Men—n
		5	5r	5	5p	

Note: Designation should be made in upper or lower left-hand corner of quantity column square.

No.	Panelboards						Raceway													Length of Run						Outlets									Wire or Cable						Misc.	
	Size	Fl.	Surf.	Type Const.	Wood Base	Size	Type	Emb'd	Exp.	Furred	Frame	Hanger	Mas'y	Ells	Fit.	Boxes	Term.	Quantity	Type	Type	Fl/Surf.	Amps	Volts	Poles.	Size	Type	No Wires	No./Pair	Length of Pull	Term.	Terminal Block

NAME	**SPECIAL SYSTEMS TAKEOFF SCHEDULE**
NUMBER	**E-13**
ORIGINATOR	Estimator
COPIES	2
DISTRIBUTION	Estimator, Job Processor
SIZE	17 × 11 in. (Copy form at 170% of original.)
PURPOSE	This schedule documents the takeoff information relative to special systems equipment, control, raceway, wire, and outlets:

1. *Special systems* consist of sound system, annunciator systems, music broadcast systems, alarm systems, and signal systems.
2. The first section of the form provides for a description and physical data of the equipment used by the system, showing the dimensions, weight, and mounting details.
3. The next section is for recording the quantities of raceway and fittings. The even footage length is posted in the appropriate column of one of the four types of construction.
4. The third section provides for recording quantities of or cable required to make the installation.
5. The fourth section provides for recording the quantity and details of outlets required. In most cases the outlets are junction boxes with blank covers.

Equipment listed on this form is posted to the Pricing Sheet (Form E-19), Equipment Work Schedule.

Raceway and fittings are posted to the Branch Circuit Tabulating Sheet (Form E-16).

Wire or cable is posted to the Conductor Tabulating Sheet (Form E-17).

Outlets should be detailed on Form E-10 and posted on the Branch Circuit Tabulating Sheet (Form E-16) and the Finishing Tabulating Sheet (Form E-18).

SPECIAL SYSTEMS TAKEOFF SCHEDULE

Job Name: _____

Sheet Number: ____ of ____
Floor Number: ____ Building: ____
Estimator: _____
Date: _____

Signals Used on Forms for Flagging Deviations from Standards

Height of Installation

Height	Designation
15–25 ft	o
26–35 ft	x
36–50 ft	✓

Conduit Parallel Runs

Number	Designation
2	2r
3	3r
4	4r
5	5r

Wire Pulls in Parallel

Number	Designation
2	2p
3	3p
4	4p
5	5p

Miscellaneous Signals

Density of items—d

Number of Men—n

Note: Designation should be made in upper or lower left-hand corner of quantity column square.

Type System	Desc.	Equipment				Raceway										Wire or Cable							Outlets					
		Size	Wgt.	Mount.	Wire Term.	Size	Type	Length of Run				Ells	Fit.	Term.	Pull Boxes	Size	Type	Length Pull	No./Wire	No./Pairs	Term	Term. Block	Quan.	Type	Fl.S.f.	Amps	Volts	Poles
								Emb'd	Expos'd	Frame	Hanger																	

	NAME	**EMBEDDED TABULATION SHEET**

NAME **EMBEDDED TABULATION SHEET**

NUMBER **E-14**

ORIGINATOR **Estimator**

COPIES **2**

DISTRIBUTION **Estimator, Job Processor**

SIZE **17 × 11 in. (Copy form at 170% of original.)**

PURPOSE The purpose of this form is to tabulate and summarize the recorded take-off quantities of embedded conduit and fittings, with the ultimate objective of producing a bill of material for all embedded materials.

On this form, accumulate the quantities from different takeoff forms of like materials installed under similar construction conditions.

Items of material posted and summarized on this form are derived from a number of takeoff forms:

1. Form E-3, Service, Metering, and Grounding Takeoff Sheet
2. Form E-7, Feeder and Busway Takeoff Schedule
3. Form E-8, Branch Circuit Takeoff Sheet (or the shortcut sheet, Form E-9A)
4. Form E-11, Power System Takeoff Schedule
5. Form E-12, Telephone System Takeoff Schedule
6. Form E-13, Special Systems Takeoff Schedule

Provisions are made in a section of the form to list boxes, fittings, sleeves, and trenching.

Separate sheets of this form are used for each type of conduit.

Many identical items of material appear on the various takeoff forms. It is essential to accumulate them in order to:

1. Accumulate identical items of material in the same type of construction and at the same height from the floor.
2. Provide a system transition from takeoff to construction Work Schedule.
3. Furnish a summary of quantities of identical materials for listing on pricing sheets according to the construction Work Schedule.
4. Eliminate duplicate listing and pricing.
5. Supply a vehicle for listing items of material on Bills of Material by the construction Work Schedule for purposes of scheduling material to the job for different stages of electrical construction.

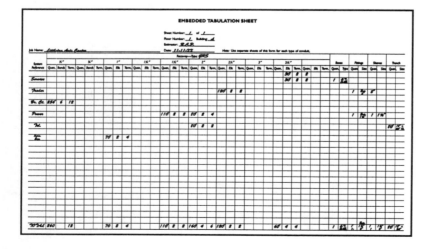

EMBEDDED TABULATION SHEET

Sheet Number: _____ of _____
Floor Number: _____ Building: _____
Estimator: _____
Date: _____

Job Name: _____

Note: Use separate sheets of this form for each type of conduit.

System Reference	Raceway—Type:																									Boxes			Fittings		Sleeves		Trench	
	½"			¾"			1"			1¼"			1½"			2"			2½"			3"			3½"										
	Quan.	Bends	Term.	Quan.	Bends	Term.	Quan.	Ells	Term.	Quan.	Ells	Term.	Quan.	Ells	Term.	Quan.	Ells	Term.	Quan.	Ells	Term.	Quan.	Ells	Term.	Quan.	Ells	Term.	Quan.	Type	Quan.	Size	Quan.	Size	Quan.	Size

NAME	**FEEDER CONDUIT TABULATING SHEET**	
NUMBER	**E-15**	
ORIGINATOR	**Estimator**	
COPIES	**2**	
DISTRIBUTION	**Estimator, Job Processor**	
SIZE	**17 × 11 in. (Copy form at 170% of original.)**	

PURPOSE On this form, tabulate and summarize the recorded quantities of identical materials of the feeder conduit system, as shown on different take-off sheets, which are installed under similar construction conditions. The ultimate objective is to produce a Bill of Material for feeder conduit. One of the key elements in pricing the labor required to install them is the height at which they are installed.

Form E-15 is designed to accommodate conduits installed at different heights:

	Factor
1. Standard conditions for a ceiling height of 14 ft.	0
2. Height of 15–25 ft, designated by a dot (•)	1.50
3. Height of 26–35 ft, designated by ×	1.70
4. Height of 36–50 ft, designated by an open dot (○)	2.00

In addition, each of the conduit runs is designated as being installed in one of two types of construction:

a. Exposed

b. On hangers

5. Conduit fitting labor units are affected by height to the same extent as raceway.

Each column on the form is summarized, and the totals are posted on the bottom line.

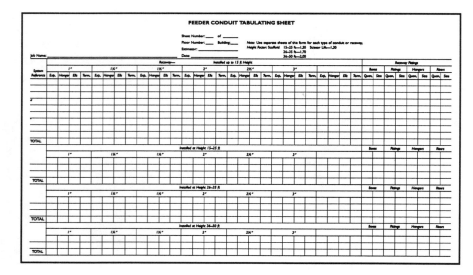

All of the totals in each section are posted separately on the Pricing Sheet (Form E-19), Feeder Work Schedule.

Tabulating forms are used for consolidating the quantities of these items, and Form E-15 serves that purpose for feeder conduits (which are not embedded). As has been noted, Form E-7 provides a method of designating height at the time of takeoff, and Form E-15 is set up to facilitate the transfer of the takeoff information inasmuch as it is subdivided by height.

FEEDER CONDUIT TABULATING SHEET

Sheet Number:____ of ________

Floor Number:____ Building:____

Estimator: ________________

Date: ________________

Job Name: ________________________________

Note: Use separate sheets of this form for each type of conduit or raceway.
Height Factor: Scaffold 15–25 ft—1.50 Scissor Lift—1.30
26–35 ft—1.70
36–50 ft—2.00

	Raceway— Installed up to 15 ft Height																												Raceway Fittings							
	1″				1¼″				1½″				2″				2½″				3″								Boxes		Fittings		Hangers		Risers	
System Reference	Exp.	Hanger	Ells	Term.	Exp.	Hanger	Ells	Term.	Exp.	Hanger	Ells	Term.	Exp.	Hanger	Ells	Term.	Exp.	Hanger	Ells	Term.	Exp.	Hanger	Ells	Term.	Exp.	Hanger	Ells	Term.	Quan.	Size	Quan.	Size	Quan.	Size	Quan.	Size
TOTAL																																				

	Installed at Height 15–25 ft																																			
	1″				1¼″				1½″				2″				2½″				3″								Boxes		Fittings		Hangers		Risers	
TOTAL																																				

	Installed at Height 26–35 ft																																			
	1″				1¼″				1½″				2″				2½″				3″								Boxes		Fittings		Hangers		Risers	
TOTAL																																				

	Installed at Height 36–50 ft																																			
	1″				1¼″				1½″				2″				2½″				3″								Boxes		Fittings		Hangers		Risers	
TOTAL																																				

NAME **BRANCH CIRCUIT TABULATING SHEET**

NUMBER **E-16**

ORIGINATOR **Estimator**

COPIES **2**

DISTRIBUTION **Estimator, Job Processor**

SIZE **17 × 11 in. (Copy form at 170% of original.)**

PURPOSE Tabulate and summarize the quantities of materials of similar type of the branch circuit system, as shown on takeoff forms E-8, E-9A, E-10, E-11, E-12 and E-13, that are installed under similar construction conditions. The objective is to produce a Bill of Material for branch circuit material.

There are eight sections on this form on which to accumulate material quantities from other takeoff forms:

1. Raceway: A separate copy of Form E-16 is required for each type of raceway.
2. Nonmetallic cable
3. Outlet boxes
4. Plaster covers
5. Box hangers
6. EMT fittings
7. GRS fittings
8. Cable fittings

There is space for three sizes of EMT, and under each size there are columns for three types of construction.

The quantity of each item is totaled at the foot of each column.

Totals of material items are transferred from this sheet to the Pricing Sheet (Form E-19), Branch Circuit Work Schedule.

BRANCH CIRCUIT TABULATING SHEET

Sheet Number: ____ of ________

Floor Number: ____ Building: ____

Estimator: ____________________

Job Name: ______________________________ Date: ____________________

System	Raceway—Type:										Cable/Ground Type						Boxes, Outlet										Covers, Plaster					
	½ in.			¾ in.			1 in.				Number/Condition						4 in. sq.		4 in. sq.	4 11/16 in. sq.		4 in. sq.		Sec. Sw.	Conc. Ring	Sing. Fl. Bx.	1 Gang		2 Gang		Out. Cov.	Conc. Cov.
	Exposed	Frame	Furred	Exposed	Frame	Furred	Exposed	Frame	Furred	Ells	12-2	12-3			3-3	Term.	1½" D.	2 1/8" D.	Bracket	1½" D.	2 1/8" D.	1½" D.	2 1/8" D.				½"	¾"	½"	¾"		
TOTALS																																

System	Box Hangers							Fittings: E.M.T.												Fittings: GRS									Fittings, Cable			
	Exp.	Str.	Deep	SH	Jack	Deep	CAD	Connectors			Couplings			Clamps, 1 hole			Straps, 2h.			Locknuts			Bushings			2h. straps			St.	Box		
	18"	18"	18"	18"		26"	MSC	½"	¾"	1"	½"	¾"	1"	½"	¾"	1"	½"	¾"	1"	½"	¾"	1"	½"	¾"	1"	½"	¾"	1"	Nails	Conn.	Nails	Straps
TOTALS																																

NAME	**CONDUCTOR TABULATING SHEET**
NUMBER	**E-17**
ORIGINATOR	**Estimator**
COPIES	**2**
DISTRIBUTION	**Estimator, Job Processor**
SIZE	**17 × 11 in. (Copy form at 150% of original.)**
PURPOSE	On this form, tabulate and summarize the recorded quantities of wire and cable for the purpose of accumulating totals from different takeoff sheets of like materials installed under similar conditions of length of run and number of conductors. The objective is to produce a Bill of Material.

1. The first section is divided into two identical sections, which are provided to accumulate the totals of branch circuit wire according to:
   a. The sizes 12, 10, and 8.
   b. The number of conductors in the run, such as 2, 3, 4, or 5.
   Totals are also accumulated for wire connectors.
2. Four sections are provided to accumulate the totals of feeder wire and connectors according to the length of run:
   a. Runs of not over 50 ft
   b. Runs of not over 100 ft
   c. Runs of not over 150 ft
   d. Runs over 150 ft
3. Columns are provided for 3- and 4-wire runs, but 2-wire runs should also be recorded separately.
4. The individual length of run is recorded separately in the appropriate section according to length, in the appropriate column according to wire size, and in the appropriate column according to the number of conductors.

   Each column is then summarized by length of run (including make-up) and converted to total footage of a single conductor by multiplying the length of run by the number of conductors. Bear in mind that there is a separate labor unit for each wire size according to the number of conductors and length of run.
5. Post the quantity of connectors according to size and type at the end of each section.
6. When more than one type of wire is involved in the takeoffs, use a different sheet of Form E-17 for each type of wire.

All the information on this form is posted on the Pricing Sheet (Form E-19), Conductor Work Schedule.

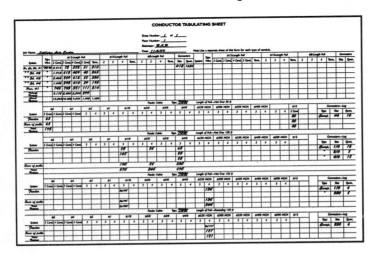

CONDUCTOR TABULATING SHEET

Sheet Number: ____ of ________

Floor Number: ____________

Estimator: ____________

Job Name: ______________________________ Date: ____________

Note: Use a separate sheet of this form for each type of conduit.

System	Type Wire	#12-Length Pull					#10-Length Pull				#8-Length Pull				Connectors		System	Type Wire	#12-Length Pull					#10-Length Pull				#8-Length Pull				Connectors	
		2 Cond.	3 Cond.	4 Cond.	5 Cond.	Term.	2	3	4	Term.	2	3	4	Term.	Size	Quan.			2 Cond.	3 Cond.	4 Cond.	5 Cond.	Term.	2	3	4	Term.	2	3	4	Term.	Size	Quan.

Feeder Cable: Type ______ Length of Pull—Not Over 50 ft

System	#6		#4		#2		#1		#1/0		#2/0		#3/0		#4/0		#250 MCM		#300 MCM		#350 MCM		#400 MCM				#16			Connectors—Lug		
	3 Cond.	4 Cond.	3 Cond.	4 Cond.	3	4	3	4	3	4	3	4	3	4	3	4	3	4	3	4	3	4	3	4	3	4				Type	Size	Quan.
Total																																

Feeder Cable: Type ______ Length of Pull—Not Over 100 ft

System	#6		#4		#2		#1		#1/0		#2/0		#3/0		#4/0		#250 MCM		#300 MCM		#350 MCM		#400 MCM		#500 MCM		#16			Connectors—Lug		
	3 Cond.	4 Cond.	3 Cond.	4 Cond.	3	4	3	4	3	4	3	4	3	4	3	4	3	4	3	4	3	4	3	4	3	4				Type	Size	Quan.
Total																																

Feeder Cable: Type ______ Length of Pull—Not Over 150 ft

System	#6		#4		#2		#1		#1/0		#2/0		#3/0		#4/0		#250 MCM		#300 MCM		#350 MCM		#400 MCM		#500 MCM		#16			Connectors—Lug		
	3 Cond.	4 Cond.	3 Cond.	4 Cond.	3	4	3	4	3	4	3	4	3	4	3	4	3	4	3	4	3	4	3	4	3	4				Type	Size	Quan.
Total																																

Feeder Cable: Type ______ Length of Pull—Exceeding 150 ft

System	#6		#4		#2		#1		#1/0		#2/0		#3/0		#4/0		#250 MCM		#300 MCM		#350 MCM		#400 MCM		#500 MCM		#16			Connectors—Lug		
	3 Cond.	4 Cond.	3 Cond.	4 Cond.	3	4	3	4	3	4	3	4	3	4	3	4	3	4	3	4	3	4	3	4	3	4				Type	Size	Quan.
Total																																

FINISHING TABULATING SHEET

E-18

Estimator

2

Estimator, Job Processor

8½ × 11 in.

PURPOSE Tabulate and summarize the recorded quantities of identical items of material used in the finishing stages of the electrical installation work. The objective is to produce a Bill of Material for the finishing stages.

1. The first section of the form is devoted to recording the quantities of devices.
2. The next section is devoted to recording the quantities of outlet plates.
3. The third section of the form is devoted to recording the quantities of motor control outlet switches, thermostats, and similar temperature and heating control devices and switches, together with associated items of equipment.
4. The fourth section is devoted to portable cords and attachment caps.

All the items on this form are summarized and posted on the Pricing Sheet (Form E-19), Finishing Work Schedule.

FINISHING TABULATING SHEET

Floor Number: _1_ Building: _A_
Estimator: _R.C.P_
Job Name: _Littleton Arts Center_   Date: _1-10-XX_

Plates—Finish: Ivory

System	1G. Dup. Recep.	2G. Dup. Recep.	2G. Comb. Sw. & Rec.	1G. Sing. Recep.			1G. Tumb. Switch	2G. Tumb. Switch	3G. Tumb. Switch		1G Tel. Plate	1G. Blank
Br. Ct.	84						48	12	4			
Power				6								
Tel.											6	
Total	84			6			48	12	4		6	

Receptacles—Type: Finish: Switches—Type: Spec. Fin.: Iv.

System	Duplex	Single	1-φ 3-Wire 10A	1-φ 3-Wire 20A	3-φ 4-Wire 20A	3-φ 4-Wire 30A	50A	Clock	Plas. Prot.	1-Pole	2-Pole	3-Way	4-Way	Photo Cell
Br. Ct.	84								168	54		8	2	
Power			4		1			4						
Total	84		4		1			4	168	54		8	2	

Cord Caps								Cord—Type:					
System	10A 3W	20A 3W	20A 4W	30A 3W	30A 4W	50A 3W	50A 4W	18-2	18-3	18-4	16-2	16-3	16-4
Br. Ct.													
Power	4		1										
Total	4		1										

System	Push But. Control Quan.	Size	Type	Cont. Trans. Quan.	Size	Type	Thermostat Quan.	Size	Type	U.F.D. Risers Quan.	Size	Type	T.O. Switches Quan.	Size	Type
Br. Ct.				1	50VA	M	4	Reg.	G.P.	—			—		
Pwr	4	S/S	G.P.												
Total	4	S/S	G.P.	1	50VA	M	4	Reg.	G.P.						

FINISHING TABULATING SHEET

Floor Number: _____ Building: ___

Estimator: _____

Job Name: _____

Date: _____

System	| Plates—Finish:												
	1G. Dup. Recep.	2G. Dup. Recep.	2G. Comb. Sw. & Rec.	1G. Sing. Recep.			1G. Tumb. Switch	2G. Tumb. Switch	3G. Tumb. Switch		1G Tel. Plate	1G. Blank	
Total													

System	Receptacles—Type:		Finish:							Switches—Type:			Fin.:		
	Duplex	Single	1-φ 3-Wire		3-φ 4-Wire			Clock	Plas. Prot.	1-Pole	2-Pole	3-Way	4-Way		Photo Cell
			10A	20A	20A	30A	50A								
Total															

System	Cord Caps							Cord—Type:					
	10A 3W	20A 3W	20A 4W	30A 3W	30A 4W	50A 3W	50A 4W	18-2	18-3	18-4	16-2	16-3	16-4
Total													

System	Push But. Control			Cont. Trans.			Thermostat			U.F.D. Risers			T.O. Switches		
	Quan.	Size	Type	Quan.	Size	Type	Quan.	Size	Type	Quan.	Size	Type	Quan.	Size	Type
Total															

NAME	**PRICING SHEET**
NUMBER	**E-19**
ORIGINATOR	**Estimator**
COPIES	**3**
DISTRIBUTION	**Contract Manager, Job Processor, Estimator**
SIZE	**17 × 11 in. (Copy form at 170% of original.)**

PURPOSE   On this form, list, price, labor, factor, schedule, and summarize by Work Schedule all the items of material prepared by the estimator. The purpose is to produce a Bill of Material for each work schedule.

This form serves multiple purposes. On small jobs everything is listed without separation for scheduling. On larger jobs, where it is desirable to separate functions for costing and scheduling, this form serves as seven forms because it is divided into seven Work Schedules:

1. Embedded
2. Feeder
3. Equipment
4. Branch Circuit
5. Conductor
6. Lighting Fixtures
7. Finishing

The sequence of this listing corresponds to the sequence in which electrical work is done in the field according to the progress of construction. The material listed on this form for scheduling is either transferred directly from a takeoff form or from the five tabulating forms. All items are listed so that pricing, laboring, and factoring can be properly applied, and each sheet is totaled separately.

List material items by name first, then by the size or type in numerical sequence. As previously mentioned, it is also necessary to separate the same item by type of construction, runs, parallel pulls, etc. For example:

1. Footages of #12 wire must be listed according to the number of wires in the run, such as 2, 3, 4, 5, and so on.
2. Feeder wire listings must recognize the length of the run, number of conductors, parallel pulls, and the like.
3. Conduit listings must recognize height factor, type of construction, and parallel runs.

In all cases, only standard labor units in man-hours are used for posting in the appropriate column under labor cost. Deviations are applied as a multiplying factor to the standard unit.

Separate sheets of this form should be numbered in consecutive order, such as P-1, P-2, etc., but the sheets for each Work Schedule should be kept together in sequence.

All Pricing Sheets are individually totaled for material cost and labor man-hours, and they are posted individually by Work Schedule or alternate on the Recapitulation Summary Sheet (Form E-20).

PRICING SHEET

Sheet Number: _1_ of _1_
Work Schedule: _Conductor_
Estimator: _R.C.P._
Job Name: _Littleton Arts Center_ Date: _11-10-XX_

Factor reference tables (top of sheet):

Length/Pull	Multiplier	Pulls in Parallel	Multiplier	Height Factor		Multiplier	Conduit Parallel Runs	Multiplier	Density Factor	Number of Mult. Factors
50 ft	1.50	1	1.00	Scaffold	15'-25'	1.50	2	0.85	0.70	1.50
100 ft	1.00	2	0.90	"	26'-35'	1.70	3	0.80	to	to
150 ft	0.85	3	0.85	"	35'-50'	2.00	4	0.77	0.87	2.50
200 ft	0.70	4	0.80	Ladder	20'-22'	1.70	5	0.74		
300 ft	0.60	5	0.78	Scissor	15'-30'	1.50				

Ref. No.	Description	Quantity	Unit	Per	Extension	Per-Foot Labor Unit Man-Hrs.	No. of Cond's	Length of Pull	Len/Pull Factor	Parallel Pulls	P.P. Factor	Factored Labor Unit	Labor Man-Hours
E-17	Wire #12 THW Copper	13,356	33.60	M	616.76	.0086	2	—	—	—	—	.0086	157.86
"	" " " "	9,426	"	"	316.71	.0065	3	—	—	—	—	.0065	61.27
"	" " " "	9,576	"	"	321.75	.0060	4	—	—	—	—	.0060	57.46
"	" " " "	1,495	"	"	50.25	.0055	5	—	—	—	—	.0055	8.22
E-17	Wire #6 THW Copper	40	137.00	M	5.48	.0167	2	20	1.50	—	—	.025	1.00
"	" " " "	195	"	"	26.71	.0125	3	65	1.50	—	—	.0187	3.65
E-17	Wire #1/0 THW Cop. (2 Pulls)	570	455.00	M	259.35	.0244	3	95	1.00	2	0.90	.0220	12.54
"	" " " "	438	455.00	M	199.29	.0244	3	146	0.85	—	—	.0207	9.07
E-17	Wire #2/0 THW Copper	344	560.00	M	192.64	.0250	4	86	1.00	—	—	.0250	8.60
E-17	Wire #4/0 THW Cop. (3 Pulls)	690	870.00	M	600.30	.0395	3	77	1.00	3	0.85	.0335	23.12
E-17	Wire #250 MCM THW Cop.	544	1065.00	M	579.36	.0382	4	136	0.85	—	—	.0325	17.68
"	" " " " (2 Pulls)	724	"	M	771.06	.0382	4	181	0.70	2	0.90	.024	17.38
E-17	Conn., Scotchlok 16/10 Cap.	1634	3.15	C	51.47	.03 ea.							49.02
"	" " 2/6 Cap.	10	12.30	C	1.23	.04 ea.							0.40
E-17	Lug. Comp. Type #1/0 Cap.	18	87.13	C	15.68	.23 ea.							4.14
"	" " #2/0 "	8	102.88	C	8.20	.25 ea.							2.00
"	" " #4/0 "	18	138.50	C	24.93	.30 ea.							5.40
"	" " #250 "	14	161.13	C	22.56	.37 ea.							5.18

Total Material Cost (This Sheet) $4063.71

Total Labor Man-Hours (This Sheet) 443.92

This Pricing Sheet is used in this instance to summarize the Conductor Work Schedule only. Each job will have this Work Schedule and six others, all using this one type of form (E-19) for a Recapitulation Summary Sheet (Form E-20).

PRICING SHEET

Length/Pull	Multiplier		Pulls in Parallel	Multiplier		Height Factor	Multiplier		Cond. Panel Run Multiplier		Conduit Parallel Runs	Multiplier		Density Factor 0.70 to 0.87	Number of Mult. Factors 1.50 to 2.50
50 ft	1.50		1	1.00		Scaffold 15'25'	1.00		1.50		2	0.85			
100 ft	1.00		2	0.90		" 26'-35'	0.90		1.70		3	0.80			
150 ft	0.85		3	0.85		" 35'-50'	0.85		2.00		4	0.77			
200 ft	0.70		4	0.80		Ladder 20'-22'	0.80		1.70		5	0.74			
300 ft	0.60		5	0.78		Scissor 15'-30'	0.78		1.50						

Conduit Factors: Embed'd 1.30 | Exposed 1.15 | Furred Hanger Frame 1.15 | Wood Frame 1.10 | 1.45

Standard Labor Unit Based on Conditions of Installation and Usage

Per-Foot Labor Unit Man-Hrs.

Conductor Factors: No. of Cond's | Length of Pull | Len/Pull Factor | Parallel Pulls | P.P. Factor

Conduit Factors: Install'd Height | Height Factor

Cond. Panel Run: Number in Par'l | P/R. Factor

Density: No. in Density | Density Factor

No./Workers: Excess Work | N/M Factor

Factored Labor Unit

Labor Man-Hours

Ref. No.	Description	Quantity	Material Cost Unit	Per	Extension

Total Material Cost (This Sheet) **$4063.71**

Total Labor Man-Hours (This Sheet) **443.92**

NAME	**RECAPITULATION SUMMARY SHEET**
NUMBER	**E-20**
ORIGINATOR	**Estimator**
COPIES	**3**
DISTRIBUTION	**Contract Manager, Job Processor, Estimator**
SIZE	**17 × 11 in. (Copy form at 170% of original.)**
PURPOSE	On this form, tabulate and summarize the material cost and labor hours for the Work Schedules and bid alternates that have been totaled on several Pricing Sheets (Form E-19). The totals from each sheet are listed separately on this Form E-20, and the material cost (adjusted) and total labor hours are posted on the Bid Summary Sheet (Form E-21). These two cost items from Form E-20 are posted on Form E-21. The estimator can then proceed with the Bid Summary, in order to determine the Job Margin Percentage, which provides information required to complete the Recapitulation Summary Sheet (Form E-20) for total price.

In addition, this form provides:

1. A means for applying job margins to cost to develop a price for bid alternates.
2. Basic information for scheduling material delivery to the job according to requirement dates established by the preparation of the Activity Schedule.
3. Information for job cost control, which results from comparisons of field time reports with Work Schedule totals.
4. Accurate basic information for monthly billing to customers, according to the progress of the job, which is compared with the estimated percentage of Work Schedules.

The first section of the form relates to material cost and pricing. Errata adjustments are made to the basic cost, and the adjusted material cost is then projected by the amount of the job margin, which is taken from the Bid Summary Sheet (Form E-21). This results in a total material price that is directly related to the bid price.

The second section relates to labor pricing. Errata adjustments are made to the man-hours for each Work Schedule or bid alternate. Adjusted man-hours are then projected by job factor man-hours and nonproductive foreman man-hours to result in a total quantity of man-hours for each Work Schedule. Then the average cost per man-hour for each man-hour is taken from Form E-21, and it is multiplied by the work schedule man-hours to yield the basic labor cost for each Work Schedule. Then the job margin is applied to the man-hours for each Work Schedule, and a total labor price is obtained for each Work Schedule and/or bid alternate that relates directly to the bid price.

When the bid alternates are accepted and become part of the job, it is then necessary to process them into the Work Schedules by reworking Form E-20.

The six areas across the top of the form (below the headings) are used to calculate information required for the data in the body of the form.

RECAPITULATION SUMMARY SHEET

Sheet Number: _1_ of _1_

Floor Number: _1_ Building:____

Estimator: _R.C.P._

Date: _11-9-XX_

Job Name: _Littleton Arts Center_

Note: This form is used in conjunction with two other forms to provide complete information:
1. Form E-21, Bid Summary Sheet.
2. Form E-24, Errata Sheet.

Preliminary Calculations

Estimated Cost of Mtl. and Labor:	594.379	Bid Quote:	856.722	Estimated Total M-H:	12.784	Job Factor M-H:	1.686
Less: Errata Adj.:	5.950	Less: Adj. Cost:	588.429	Less: Errata Adj.:	136	Adj. Basic M-H:	12.648
Adjusted Cost of Material and Labor:	588.429	Job Margin:	268.293	Adjusted Man-Hours:	12.648	Percent Job Fac. M-H:	13.33
		Percent Margin:	45.59				

Total Supv.-App. M-H:	3.720	Estimated Labor Cost:	314.480
Less: Prod. Supv.-App. M-H:	2.226	Less: Errata Adjustment:	2.720
Nonprod. Supv.-App. M-H:	1.494	Adjusted Labor Cost:	311.760
Percent Nonprod. M-H:	11.81	Adjusted Cost Per M-H:	$19.53

Ref. No.	Work Schedule	Material Price Data								Labor Price Data												Total Mtl. & Lab. Sch. Price
		Material Cost	Total Schedule Mtl. Cost	Errata Adj.	Adjusted Material Cost	Percent Job Margin	Amount Margin	Material Schedule Total Price	Basic Man-Hours	Total Schedule Man-Hours	Errata Adj.	Adjusted Man-Hrs.	%	Job Factor Man-Hr.	%	Nonprod. Man-Hr.	Total Schedule Man-Hrs.	Cost Per Man-Hr.	Labor Cost	% Job Margin	Amount Margin	Labor Schedule Total Price	
PS-1	Embedded	5.506							390														
"	"	2.728	8.234	—	8.234	.456	3.755	11.989	112	502	—	502	.1333	67	.1181	59	628	$19.53	12.265	.456	5593	17.858	29.847
PS-2	Feeder	22.402							355														
"	"	9.422							1.240														
"	"	19.280	51.104	—	51.104	"	23.303	74.407	1.335	2960	—	2960	"	395	"	350	3.705	"	72.359	"	32.996	105.355	179.768
PS-3	Equipment	58.800	58.800	-1.400	57.400	"	26.174	83.574	1.800	1.800	-40	1.760	"	235	"	208	2.203	"	43.025	"	19.619	62.644	146.218
PS-4	Branch Circuit	3.920							251														
"	"	3.742							190														
"	"	1.450	9.112	—	9.112	"	4.155	13.267	196	617	—	617	"	82	"	73	772	"	15.077	"	6.875	21.952	35.219
PS-5	Conductor	8.128							455														
"	"	35.932	44.060	-830	43.230	"	19.713	62.943	1345	1.800	-26	1.774	"	236	"	209	2.219	"	43.337	"	19.762	63.099	126.042
PS-6	Lighting Fix.	70.000	70.000	-1.000	69.000	"	31.464	100.464	2.900	2.900	-70	2.830	"	377	"	334	3541	"	69.156	"	31.535	100.691	201.155
PS-7	Finishing	2.280	2.280	—	2.280	"	1.040	3.320	180	180	—	180	"	16	"	14	150	"	2.930	"	1.336	4.266	7.586
PS-8	Alternate #1	7.400							405														
"	"	9.300							302														
"	"	3.880	20.580	—	20.580	"	9.384	29.964	463	1.170	—	1.170	"	156	"	138	1464	"	28.592	"	13.038	41.630	71.594
PS-9	Alternate #2	15.729	15.729	—	15.729	"	7.684	22.993	915	915	—	915	"	122	"	108	1145	"	22.362	"	10.197	32.559	55.552
																Adj.	137				Adj.	3.747

Total Material Cost: 279.899 | **-3830** | 276.669 | **Total Material Price:** 402.921 | **Total Basic Man-Hours:** 12.784 | -136 | 12.648 | **Total Adj. Job Man-Hours:** -15.964 | 309.103 | **Total Labor Price:** 450.054 | 856.722

RECAPITULATION SUMMARY SHEET

Job Name: _____

Sheet Number: _____ of _____
Floor Number: _____ Building: _____
Estimator: _____
Date: _____

Note: This form is used in conjunction with two other forms to provide complete information:
1. Form E-21, Bid Summary Sheet.
2. Form E-24, Errata Sheet.

Preliminary Calculations

Estimated Cost of Mtl. and Labor:	_____	Bid Quote:	_____	Estimated Total M-H:	_____	Job Factor M-H:	_____	Total Supv.-App. M-H:	_____	Estimated Labor Cost:	_____
Less: Errata Adj.:	_____	Less: Adj. Cost:	_____	Less: Errata Adj.:	_____	Adj. Basic M-H:	_____	Less: Prod. Supv.-App. M-H:	_____	Less: Errata Adjustment:	_____
Adjusted Cost of Material and Labor:	_____	Job Margin:	_____	Adjusted Man-Hours:	_____	Percent Job Fac. M-H:	_____	Nonprod. Supv.-App. M-H:	_____	Adjusted Labor Cost:	_____
		Percent Margin:	_____					Percent Nonprod. M-H:	_____	Adjusted Cost Per M-H:	_____

Material Price Data

Ref. No.	Work Schedule	Material Cost	Total Schedule Mtl. Cost	Errata Adj.	Adjusted Material Cost	Percent Job Margin	Amount Margin	Material Schedule Total Price
Total Material Cost:						Total Material Price:		

Labor Price Data

Basic Man-Hours	Total Schedule Man-Hours	Errata Adj.	Adjusted Man-Hrs.	%	Job Factor Man-Hr.	%	Nonprod. Man-Hr.	Total Schedule Man-Hrs.	Cost Per Man-Hr.	Labor Cost	% Job Margin	Amount Margin	Labor Schedule Total Price	Total Mtl. & Lab. Sch. Price
Total Basic Man-Hours:			Total Adj. Job Man-Hours:							Total Labor Price:				

NAME	**BID SUMMARY SHEET**
NUMBER	**E-21**
ORIGINATOR	Estimator
COPIES	3
DISTRIBUTION	Contract Manager, Job Processor, Estimator
SIZE	8½ × 11 in.

PURPOSE   This form documents all the costs of the estimate, with properly evaluated overhead and profit margins. The purpose is to arrive at a price to submit for bidding purposes.

The form provides several sections in which orderly and consecutive determinations are made to arrive at a competitive bid price.

The cost of material and man-hours of labor are posted from the Recapitulation Summary Sheet (Form E-20).

A separate form is provided for the evaluation of job factor man-hours of labor— Job Factor Evaluation Sheet (Form E-22). The determinations are made on Form E-22 and posted on the proper line on the Bid Summary Sheet (Form E-21) to arrive at a total of basic labor man-hours.

Total labor hours are determined on the Manpower Chart (Form E-23), and they are transferred to the proper line on the Bid Summary Sheet (Form E-21). The chart enables the estimator to determine the man-hours of each category of labor, such as:

1. General Foreman (productive and nonproductive)
2. Foreman (productive and nonproductive)
3. Journeyman
4. Apprentice (all classes, productive and nonproductive)

Direct job expenses are determined by the estimator and posted on the proper lines of Form E-21.

Overhead expense represents the portion of the total cost of doing business that this job will bear. Each job must bear its proper share of the total cost. The best way to determine this is to prorate it according to man-hours, or labor burden. From the business budget is determined the overhead cost per budgeted man-hour. For example, if this figure were $9.00/M.H. and the job under consideration involved 10,000 M.H., the allowance for overhead for the job would be $90,000.

Profit is added as a percentage of prime cost according to a chart that varies the amount by the size of the job in man-hours and by the duration of the job in months.

Certain fees are added at the foot of the estimate sheet, which are usually based on the total job price. These are included in the price with no additives.

At the last minute before bidding, price changes are generally called in; these are recorded on the Errata Sheet (Form E-24) and entered on the Bid Summary Sheet for adjusting the bottom line of the estimate.

Form E-21 furnishes four statistics used on Form E-20, Recapitulation Summary Sheet:

1. Percent job margin
2. Labor cost per man-hour
3. Percentage job factor man-hours
4. Percent nonproductive man-hours

BID SUMMARY SHEET

Sheet Number: _1_ of: _1_

Estimator: _R.C.P_

Job Name: _Littleton Arts Center_   Date: _4-20-92_

			279270—
1. Estimated Material Cost		Labor Cost	629—
Expendible Material 0.20%: _314,480_		12,784	279899—
2. Estimated Labor Man-Hours			
3. Job Factors: Height of Building			
4. Prefabrication	614		
5. Productivity _−9.0_ %	1150		
6. Similar Buildings			
7. Typical Floors			
8. Working Conditions	1150	1686	
9. Basic Labor Man-Hours		14,470	
10. Less: Productive Supervision-Apprentice Man-Hours		− 2,226	
11. Total Journeyman Man-Hours		12,244	
12. Plus: Total Supervision-Apprentice Man-Hours		3,720	
13. Total Labor Man-Hours		15,964	
14. Labor Cost: General Foreman: _880_ hr @ _24.00_		21120—	
15. Foreman _1,400_ hr @ _22.00_		31680—	
16. Journeyman _12,244_ hr @ _20.00_		244880—	
17. Apprentice _1,400_ hr @ _12.00_		16800—	314480—
18. Cost of Maerial and Labor:			594379—
19. Dir. Job Exp: Board and Lodging			
20. Equipment (Trucks, Trenchers)			
21. Freight on Material to the Job			
22. Payroll Tax: _31.4_% _314,480._	98747—		
23. Sales Tax: _6.25_% _279,899._	17494—		
24. Special Insurance			
25. Subcontracts			
26. Telephone Expense	400		
27. Trips to the Job			
28. Travel Expense	800		
29. Warehouse and Field Office			
30. Miscellaneous			117441—
31. Prime Cost:			711820—
32. Overhead Cost: _15% − 711,820_			106773—
33. Profit: _4.6% − 711,820_			32743—
34. Subtotal:			851,336—
35. Depository Fee: _0.25_% _851,336_	2128—		
36. Performance Bond: _0.60_% _851,336_	5108—		
37. Permits and Inspection:	4100—		11336—
38. Bid Price:			862672—
39. Errata Adjustment			− 5950—
40. Percent Job Margin _17.05_ Labor/M.H. _$19.70_ Material _32.6_% Labor _36.7_% Bid Quote: _856722—_			

BID SUMMARY SHEET

Sheet Number: _____ of: _____

Estimator: _____

Job Name: _____ Date: _____

1. Estimated Material Cost _____
   Expendible Material 0.20%: _____   Labor Cost _____
2. Estimated Labor Man-Hours _____ _____
3. Job Factors: Height of Building _____
4.   Prefabrication _____
5.   Productivity _____ % _____
6.   Similar Buildings _____
7.   Typical Floors _____
8.   Working Conditions _____ _____
9.   Basic Labor Man-Hours _____ _____
10. Less: Productive Supervision-Apprentice Man-Hours _____ _____
11.   Total Journeyman Man-Hours _____ _____
12. Plus: Total Supervision-Apprentice Man-Hours _____ _____
13.   Total Labor Man-Hours _____ _____
14. Labor Cost: General Foreman _____ hr @ _____ _____
15.   Foreman _____ hr @ _____ _____
16.   Journeyman _____ hr @ _____ _____
17.   Apprentice _____ hr @ _____ _____ _____
18. Cost of Maerial and Labor: .... _____
19. Dir. Job Exp: Board and Lodging _____
20.   Equipment (Trucks, Trenchers) _____
21.   Freight on Material to the Job _____
22.   Payroll Tax: _____ % _____ _____
23.   Sales Tax: _____ % _____ _____
24.   Special Insurance _____
25.   Subcontracts _____
26.   Telephone Expense _____
27.   Trips to the Job _____
28.   Travel Expense _____
29.   Warehouse and Field Office _____
30.   Miscellaneous _____ _____
31. Prime Cost: .... _____
32. Overhead Cost: _____ _____
33. Profit: _____ _____
34. Subtotal: .... _____
35. Depository Fee: _____ % _____ _____
36. Performance Bond: _____ % _____ _____
37. Permits and Inspection: _____ _____
38. Bid Price: .... _____
39. Errata Adjustment _____
40. Percent Job Margin _____ Labor/M.H. _____ Material _____ % Labor _____ % Bid Quote: _____

NAME **JOB FACTOR EVALUATION SHEET**

NUMBER **E-22**

ORIGINATOR **Estimator**

COPIES **2**

DISTRIBUTION **Estimator, Job Processor**

SIZE **11 × 8½ in.**

PURPOSE With this chart, you can determine the job factor man-hours to be added to the adjusted man-hours on the Bid Summary Sheet (Form E-21).

A *job factor* is defined as a modification to labor cost occasioned by conditions of installation that differ from the established conditions on which standard labor costs are based. Job factors are applied to deviations from overall job conditions rather than from individual labor units.

This chart shows a majority of job conditions to which factors are applied; and shows a range of minimum and maximum factors which may be applied to the man-hours that are subject to change.

The first item is used to indicate the number of floors or divisions of work involved in one of the job conditions subject to factoring. These items may represent the number of units that may occur in multiple of the unit quantity of man-hours shown in the next column. For example, an item might be a typical floor in a six-floor building; so the six floors would be posted in the column adjacent to Height of Building Floor factor. If each typical floor took 3000 man-hours, then this figure would appear in column 2.

The third column represents the product of columns 1 and 2.

The Minimum Factor Multiplier is used to determine the factor by which the man-hours involved may be multiplied to provide a minimum adder or deduction for the job condition involved.

The maximum factor column displays the maximum recommended factor by which the man-hours involved may be multiplied to provide an adder or deduction for job condition deviation.

The Multiplier Used column would be used to indicate the actual factor selected by the estimator to apply to the man-hours involved. The total adjustment for each of the job conditions is shown on the proper line in the Man-Hour Adjustment column. This same figure will also be posted on Bid Summary Sheet (Form E-21), on the line having a counterpart number.

JOB FACTOR EVALUATION SHEET
(Used in Conjunction with Bid Summary Sheet [Form E-21])

Sheet Number: *1* of *1*
Floor Number: *1* Building: *A*
Estimator: *R.C.D.*
Date: *11-7-77*

Job Name: *Littleton Arts Center*

Form E-21 Line Number	Description of Job Conditions	Number of Items	Unit Quantity of Man-Hours	Man-Hours Involved	Minimum Factor Multiplier	Maximum Factor Multiplier	Multiplier Used	Man-Hour Adjustment
3.	Height of building floor factor				+0.025	+0.30		
3.	Height of building riser factor				+0.025	+0.30		
4.	Preassembly or prefabrication			10,240	−0.05	−0.25	−0.06	−614
5.	Productivity of manpower			12,784	0.00	+0.25	+0.09	+1,150
6.	Similar buildings				−0.03	−0.07		
7.	Typical floors				−0.056	−0.106		
8.	Working conditions (Itemized below)							
a.	Continuity of work				+0.10	+0.17		
b.	Conflicts with other crafts				+0.10	+0.33		
c.	Material furnished by others				0.00	+0.06		
d.	Interruptions to work				0.00	+0.33		
e.	Rate of progress of job by general contractor			12,784	+0.05	+0.15	+0.09	+1,150
f.	Shape of space (rooms or buildings)				+0.10	+0.15		
g.	Special clothing required				+0.40	+1.00		
h.	Tooling deficiencies				0.00	+0.50		
i.	Weather or air temperature				0.00	+0.40		
j.	Work obstructions to clear				0.00	+0.40		
k.	Other							

Note: Line number refers to counterpart on Form E-21.

Total Job Factor Hours +1,684

JOB FACTOR EVALUATION SHEET
(Used in Conjunction with Bid Summary Sheet [Form E-21])

Job Name: _____

Form E-21 Line Number	Description of Job Conditions	Number of Items	Unit Quantity of Man-Hours	Man-Hours Involved	Minimum Factor Multiplier	Maximum Factor Multiplier	Multiplier Used	Man-Hour Adjustment
3.	Height of building floor factor				+0.025	+0.30		
3.	Height of building riser factor				+0.025	+0.30		
4.	Preassembly or prefabrication				−0.05	−0.25		
5.	Productivity of manpower				0.00	+0.25		
6.	Similar buildings				−0.03	−0.07		
7.	Typical floors				−0.056	−0.106		
8.	Working conditions (itemized below)							
a.	Continuity of work				+0.10	+0.17		
b.	Conflicts with other crafts				+0.10	+0.33		
c.	Material furnished by others				0.00	+0.06		
d.	Interruptions to work				0.00	+0.33		
e.	Rate of progress of job by general contractor				+0.05	+0.15		
f.	Shape of space (rooms or buildings)				+0.10	+0.15		
g.	Special clothing required				+0.40	+1.00		
h.	Tooling deficiencies				0.00	+0.50		
i.	Weather or air temperature				0.00	+0.40		
j.	Work obstructions to clear				0.00	+0.40		
k.	Other							
						Total Job Factor Hours		

Note: Line number refers to counterpart on Form E-21.

NAME	**MANPOWER CHART**
NUMBER	**E-23**
ORIGINATOR	Estimator
COPIES	1
DISTRIBUTION	Estimator
SIZE	17 × 11 in. (Copy form at 170% of original.)

PURPOSE   This form enables the estimators to project the anticipated scheduling of workers on and off the job, and to determine supervision and apprentice labor requirements. Provisions are made at the top of the form for three tabular sections in which determinations are made to provide data for use on the Bid Summary Sheet (Form E-21).

Standard labor units include the productive labor component of supervision, providing that the supervisors perform tasks that are a necessary component of a productive labor unit. These component tasks can be expressed as percentage of a productive unit:

1. Study plans and specifications: 5.0%
2. Ordering material and tools from warehouse: 2.0%
3. Receiving and storing material and tools: 4.0%

The number of workers in the crew has considerable bearing on how productive the foreman is. Under average conditions, the general foreman is 60% productive and the foreman is 45% productive when both are involved on the job. On the average, the apprentice is 75% as productive as a journeyman in terms of a standard labor unit.

After the manpower has been developed and the number of workers of the different classifications has been preliminarily determined, the supervisory and apprentice man-hours are posted in the table at the top of the graph. The productive hours are determined by applying the foregoing percentages.

The journeyman man-hours are represented by the basic man-hours shown on line 9 of the Bid Summary Sheet (Form E-21), less the productive component of supervision-apprentice man-hours. Total job man-hours now become net journeymen man-hours plus supervision-apprentice man-hours of Table 1 at the top of the form. Table 2 shows the determination of journeymen man-hours and total job man-hours.

Table 3 provides for a summary of all types of the different classes of labor in man-weeks (for verification of the Manpower Chart). The man-hours required are posted on lines 14, 15, 16, and 17 of the Bid Summary Sheet.

Since the Manpower Chart is based on man-weeks of labor, Table 3 enables the estimator to verify the original projection of man-weeks, which at this point is somewhat short of the journeyman forecast.

Therefore, the Manpower Chart may now be increased to meet the full requirements of Table 3.

The first step in preparing the Manpower Chart is to determine from the general contractor or architect the overall duration of the job. With this information, the estimator can determine the duration of the electrical work.

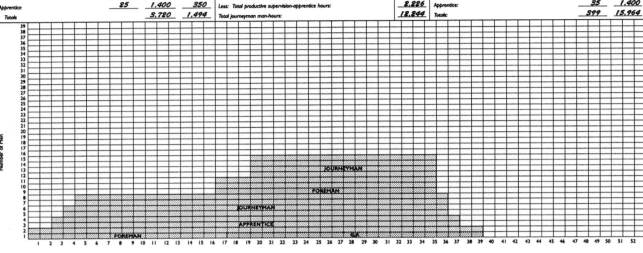

MANPOWER CHART

Sheet Number: _1_ of _1_

Floor Number: _1_ Building: _A_

Estimator: _R.C.P._

Job Name: _Littleton Arts Center_

Date: _11-14-92_

Directions:
1. Prepare chart and determine hours for general foreman, foreman, and apprentice.
2. Determine nonproductive supervision-apprentice hours.
3. Determine journeyman hours.
4. Recap. man-hours by labor classification.

Nonproductive Hours—Supervision-Apprentice

Title	Percent	Supv.-App. Hours	Nonprod. Hours
General Foreman	40	880	352
Foreman	55	1,440	792
Apprentice	25	1,400	350
Totals		3,720	1,494

Determination—Journeyman Hours

	Man-Hours
Basic Man-Hours (Line 9, E-21):	14,470
Less: Total supervision-apprentice hours: 3,720	
Nonproductive supervision-apprentice hours: 1,494	
Less: Total productive supervision-apprentice hours:	2,226
Total journeyman man-hours:	12,244

Recap. Man-Hours by Title

	Man-Weeks	Man-Hours
General foreman:	22	880
Foreman:	36	1,440
Journeyman:	306	12,244
Apprentice:	35	1,400
Totals:	399	15,964

MANPOWER CHART

Job Name: _____

Sheet Number: _____ of _____

Floor Number: _____ Building: _____

Estimator: _____

Date: _____

Directions:
1. Prepare chart and determine hours for general foreman, foreman, and apprentice.
2. Determine nonproductive supervision-apprentice hours.
3. Determine journeyman hours.
4. Recap. man-hours by labor classification.

Nonproductive Hours—Supervision-Apprentice

	Percent	Supr.-App. Hours	Nonprod. Hours
General Foreman	40		
Foreman	55		
Apprentice	25		
Totals			

Determination—Journeyman Hours

Basic Man-Hours (Line 9, E-21):

Less: Total supervision-apprentice hours:

Nonproductive supervision-apprentice hours:

Less: Total productive supervision-apprentice hours:

Total journeyman man-hours:

Recap. Man-Hours by Title

Man-Hours

	Man-Weeks	Man-Hours
General foreman:		
Foreman:		
Journeyman:		
Apprentice:		
Totals:		

Number of Men

Number of Weeks

NAME	**ERRATA SHEET**
NUMBER	**E-24**
ORIGINATOR	Estimator
COPIES	1
DISTRIBUTION	Estimator
SIZE	8½ × 11 in.

PURPOSE On this form, record last-minute changes in material prices that occur just before bid time, along with the resulting changes in labor cost, if any. The Errata Sheet enables the estimator to modify figures in the estimate, without changing the papers of the body of the estimate. All changes refer to the original quotation.

The first column provides for reference to the number of the Pricing Sheet that bore the original quotation. The description must bear a complete reference to the original quote, because this is the only written record of the transaction. Therefore, the original quote, as well as the revised quote, must be shown on this form as part of the description, and the difference is recorded in the right-hand column as to whether it is an increase or decrease.

All the items on the Errata Sheet are totaled, and a net change in material price and labor dollars is posted on the Bid Summary Sheet (Form E-21).

All changes on the Errata Sheet must be made under the proper Work Schedule or bid alternate in order to reflect them properly on the Recapitulation Summary Sheet, which is organized in this manner.

ERRATA SHEET
(Used for Making Last-Minute Changes in Bid Summary)

Sheet Number: _1_ of _1_
Estimator: _R.C.P_
Job Name: _Littleton Arts Center_   Date: _4-20-XX_

Sched. No.	Description of Revision— List by Work Schedule	Quantity Material	Quantity Labor Hours	$	Total Adjustment Add	Total Adjustment Deduct
P-3	Equipment—Orig. Quote	29,400	—			
	—Rev. "	28,000	—			1,400
P-3	" —Lower Weight		−40.	10.00		400
P-5	Conductor —Orig. Quote	22,030				
	—Rev. Quote	21,200				830
P-5	Error in est.		−26.	10.00		260
P-6	Lighting Fix.—Orig. Quote	35,000	—			
	—Rev. Quote	34,000	—			1,000
P-6	Fix. Factory Ass'y		−70.	10.00		700
	Total Labor Hours		−136			
				Totals: Add or Deduct		4,590 −
				Net Change		4,590 −

ERRATA SHEET
(Used for Making Last-Minute Changes in Bid Summary)

Sheet Number: _____ of: _____

Estimator: _____

Job Name: _____ Date: _____

Sched. No.	Description of Revision—List by Work Schedule	Quantity			Total Adjustment	
		Material	Labor Hours	$	Add	Deduct
	Total Labor Hours					

Totals: Add or Deduct

Net Change

CHAPTER SEVEN
JOB MANAGEMENT

FORM NO.	FORM NAME	PAGE
JM-1	Employee Information Record	182
JM-2	Field Personnel Time Report	184
JM-3	Termination Slip	186
JM-4	Paid Out Form	188
JM-5	Delivery Order	190
JM-6	Vacation/Sick Leave Request	192
JM-7	Memorandum	194
JM-8	Vehicle Accident Report	196
JM-9	Report of Theft, Vandalism, or Damage to Equipment, Tools, or Property	198
JM-10	Confidential Foreman Rating	200
JM-11	Revision to Bill of Material	204
JM-12	Material Scheduling Change	206
JM-13	Tool Maintenance and Repair	208
JM-14	Shop Drawing Log	210
JM-15	Shop Drawing Transmittal Form	212
JM-16	Feeder Cutting Lengths	214
JM-17	Bill of Material for Cable Field Measure	216
JM-18	Job Closeout Checklist	218
JM-19	Job Evaluation Report	220
JM-20	Estimate Evaluation Report	222
JM-21	Change Order	224
JM-22	Change Order Bid Summary	228

Contract Managers may manage as many as ten jobs, depending on the size and complexity. Their responsibility involves the following:

✓ Selection of field personnel for the jobs, including the general foreman, foreman, journeymen, and apprentices
✓ Supplying the job with material
✓ Supplying the job with tools and equipment
✓ Supplying the job with plans, specifications, and sketches, together with all information relative to the installation
✓ Handling Change Orders
✓ Acting as company liaison to the owner, architect, and general contractor
✓ Supplying the job with facilities for handling material and storing material and tools on the job
✓ Providing the accounting office with monthly information relative to the percentage of completion for billing

Next to employing competent electricians to do the work, the Contract Manager's main task is to provide material to the job in the right quantity and at the right time. Initially he must rely on the Bill of Material prepared by the processor, as well as all the other information relative to scheduling of material delivery and so forth. As the job progresses and conditions change, however, it will be necessary to modify all job documents.

The overall objective of job management is to accomplish the installation of electrical work in accordance with the terms of the contract and within the framework of the estimate. It is mandatory that the work be done within the province of the prime cost—i.e., the total cost of material, labor, and job expense. To do this, it is essential to have a plan for the effective management of the job and that plan should be disseminated to all those involved in the installation. A constant effort must be made to improve the plan, and this involves close attention to the work as the job progresses.

A major portion of the Contract Manager's time will probably be devoted to material control for the job—every job. Because the general contractor changes the type of construction from that shown on the plans, or because changes occur in the form of additions or deletions, the Contract Manager will be forced to change the Bill of Material by revision, addition, or deletion. The success, or lack of success in controlling material is manifested in three ways:

1. The amount of material ordered from the field during the job
2. The amount of material returned to the shop at the end of the job
3. The number of delays or interruptions in the actual work assignment due to lack of material or improper material (This measure of failure will probably never be known.)

If the Bills of Material were perfect, none of the aforementioned problems would occur—providing the scheduling was perfect, the Purchasing Department functioned as it should, and the suppliers met their obligations as required.

Practically all the functions of the Contract Manager involves the use of forms, for making adjustments to material lists, tool, and labor changes in order to conform to the requirements of the job as it progresses and as the construction changes. Therefore, the use of good forms to communicate these changes to the organization is essential to the success of the job.

NAME	**EMPLOYEE INFORMATION RECORD**
NUMBER	**JM-1**
ORIGINATOR	**Construction Manager**
COPIES	**2**
DISTRIBUTION	**Accounting Department, Contract Manager**
SIZE	**8½ × 11 in.**
PURPOSE	This form documents the personnel data of field employees.

This form documents the personnel data of field employees.

Each field employee is required to fill out this form so that the Payroll Department has adequate information for the company records.

This form is used for both an employment record and a termination record.

EMPLOYEE INFORMATION RECORD

New Hire:_____   Employee Number: _____

Record Change: _____   Preparation Date:_____

Termination: _____   Effective Date: _____

Name: _____ (Form W-4 must be attached for new hire.)

Address: _____

City: _____   Salary per Year: $ _____   Hourly Rate: $ _____   Change: $ _____

Termination   Retired:___   Death:___   Resigned:___   Laid off:___   Leave of Absence:___   Company Request:___

Eligible for Rehire ___   Keys/Credit Card Return:___   Uniforms:___   COBRA Insurance:___   Employee Purchase:___

Social Security Number: _____   Classification: _____   Number of Dependents: _____

_____

_____

_____

_____

_____

_____

_____

_____

_____

_____

_____

_____

_____

_____

_____

_____

_____

_____

_____

_____

_____

_____

_____

_____

_____

_____

Approvals: _____   _____   _____
             Department Manager   President   Prepared by

NAME	**FIELD PERSONNEL TIME REPORT**
NUMBER	**JM-2**
ORIGINATOR	**Field Personnel**
COPIES	**2**
DISTRIBUTION	**Payroll Department, Contract Manager**
SIZE	**11 × 8½ in.**

PURPOSE   This document records, for each individual in the field, the following information on a daily basis from a weekly Time Report:

1. Total man-hours worked on straight time, double time, or time and a half
2. Name of job and work order number for each hour
3. Code for the type of work
4. Details of the equipment used on the job, with equipment number, hours used, and mileage
5. Details of the cash expended for job expense with work order number
6. Signature of the individual

Please use pencil.

TIME REPORT

Employee
Name (Print): _____

Number: _____

Week Ending: _____

CL	Scale Rate	Job Number	Work Order Number/Code	Job Name	Monday			Tuesday			Wednesday			Thursday			Friday			Sat.			Sun.		Total Hours	T.H. Only	D.T. Only
					S.T.	T.H.	D.T.	S.T.	T.H.	D.T.	S.T.	T.H.	D.T.	S.T.	T.H.	D.T.	S.T.	T.H.	D.T.	T.H.	D.T.	D.T.				

WEEKLY TOTALS

Location In: ____ Out: ____
Location In: ____ Out: ____

Detail of Equipment Usage on the Job

Work Order Number	Description	Equipment No.	Hrs. Used	Mileage	Job Number

Job Number _____

Detail of Job Expenses

Job Number	Work Order Number	Amount

Employee Signs Here: _____

Approval _____

Use the back of form, if necessary.

NAME	**TERMINATION SLIP**
NUMBER	**JM-3**
ORIGINATOR	**Contract Manager**
COPIES	**2**
DISTRIBUTION	**Payroll Department, Contract Manger**
SIZE	**5½ × 3½ in. (Copy form at 100%; cut it out along the box rules.)**

PURPOSE This form records the dismissal of an employee.

It is to be used for all company employees, but it is especially adaptable for union employees. Undesirable workers are put on record as to their status as nonproductive.

The terminated employee must also have a Time Report completely filled out for the Payroll Department.

TERMINATION SLIP

Local Union Number: _____ Date: / _____ / _____

Employee: _____ Social Security No.: _____

Referred as: Group **I** ☐ **II** ☐ **III** ☐ **IV** ☐ Temporary Employee ☐

Employed as: _____

Terminated by:

    Quit ☐ Fired ☐ Laid off ☐ _____

Comments

_____

Eligible for rehire: ☐ **Yes** ☐ **No** Date of Eligible Rehire: _____

Approval: _____

Title: _____

NAME	**PAID OUT FORM**
NUMBER	**JM-4**
ORIGINATOR	**Field Foreman**
COPIES	**3**
DISTRIBUTION	**Contract Manager (2 copies), Foreman**
SIZE	**5½ × 8½ in. (Copy form at 100%; cut it out along the box rules.)**
PURPOSE	On this document, record the disbursement of cash made in the field on behalf of a job. It is necessary to show the work order number of the job to which the expenditure is chargeable.

PAID OUT FORM

Job Name: _____ Work Order: _____ Date: ____ / ____ / ____

Items	Account Number	Amount

Received the above amount   Total

Signed: _____

NAME	**DELIVERY ORDER**
NUMBER	**JM-5**
ORIGINATOR	**Contract Manager, Foreman, department managers**
COPIES	**3**
DISTRIBUTION	**Warehouse (2 copies), originator**
SIZE	**5½ × 8½ in. (Copy form at 100%; cut it out along the box rules.)**

PURPOSE This form records instructions to anyone in the company (to the warehouse, usually) who requires the delivery of a letter, plans, or an object to a distant location.

The deliverer must return a copy of the form to the point of origin to indicate the completion of the instructions.

The time given and the time of completion are important parts of the transaction.

DELIVERY ORDER

Date: ____ / ____ / ____   Delivery for: _____

Date Requested: _____   Time Requested: _____

Customer: _____

Number	Material Description	P.O. Number	Pick up at	Deliver to

Other Instructions: _____

_____

_____

_____

_____

_____

_____

_____

_____

_____ By: _____

NAME	**VACATION/SICK LEAVE REQUEST**
NUMBER	**JM-6**
ORIGINATOR	**Any employee**
COPIES	**3**
DISTRIBUTION	**Department manager (2 copies), individual employee**
SIZE	**5½ × 4 in. (Copy form at 100%; cut it out along the box rules.)**
PURPOSE	This form records requests from employees for time off for vacation or sick leave or for any other purpose of a useful nature. Sick leave is reported on an arrears basis because the report is made after the fact. Other requests are submitted in advance.

VACATION/SICK LEAVE REQUEST

Name: _____   Employee Number: _____
(Please Print.)

Vacation leave is requested as follows:

From: _____   To: _____   Number of Days: _____

Date Vacation Check Desired: _____ (Give one week's notice.)

Sick Leave was as follows:

From: _____   To: _____   Number of Days: _____
Approved:

_____   _____
(Department Manager)   (Signed)

Return to Personnel.

NAME	**MEMORANDUM**
NUMBER	**JM-7**
ORIGINATOR	**Any employee**
COPIES	**2**
DISTRIBUTION	**Originator, recipient**
SIZE	**5½ × 8½ in. (Copy form at 100%; cut it out along the box rules.)**
PURPOSE	With this document, one employee may communicate in an informal manner with another.
	If an answer to the communication is required, another note on a similar form would be required.

MEMORANDUM

To: _____ Date: _____

Subject: _____

Signed: _____

NAME	**VEHICLE ACCIDENT REPORT**
NUMBER	**JM-8**
ORIGINATOR	**Any truck or automobile driver**
COPIES	**1**
DISTRIBUTION	**Foreman or Contract Manager**
SIZE	**8½ × 11 in.**
PURPOSE	Use this document to report to the company office any accident involving a company vehicle or equipment.

If a serious accident or personal injury is involved, the driver of the vehicle must call the department manager or the company insurance department immediately—within 24 hours of the incident.

The following measures must also be taken:

1. Stop the vehicle immediately.
2. Assist the injured if aid is needed or requested.
3. Exchange information with other persons involved.
4. Notify the local police, sheriff, or state patrol immediately.
5. Report the accident to the State Motor Vehicle Division, in addition to the local police.

VEHICLE ACCIDENT REPORT

If serious damage or personal injury is involved (regardless of whether you think the company is liable), call your department manager or the insurance department **immediately. NEVER ADMIT LIABILITY!** The driver of any company vehicle involved in an accident **MUST** turn in this completed form to the company office within 24 hours of the accident.

Date of This Report:_____   Date of Accident:_____   Time of Accident:_____

Location of Accident:_____

(Street Address)　　　　　　(City or County)　　　　(State)

Reported to What Police Department?_____

(Attach original or copy of any information furnished by police.)

Was Traffic Citation Issued?_____   If so, What Violation and to Whom?_____

Was Accident on Private Property?　Yes:_____　No:_____　Case No.:_____

Witnesses:　(Name, Address, and Phone)　　　　　　Office's Name:_____

_____

_____

Company Vehicle

Unit Number:_____   Make, Model, Year:_____

Driver's Name:_____   Home Phone Number:_____

Home Address:_____

Driver's License Number:_____   State:_____   Date of Birth:_____

Department Where Employed:_____   Supervisor:_____

Describe Damage to Company Vehicle:_____

OTHER VEHICLE

License Plate Number:_____   State:_____   Make, Model, Year:_____

Driver's Name:_____   Home Phone No.:_____   Work Phone No.:_____

Driver's Address:_____

City:_____   State:_____   Zip:_____

Vehicle Owner's Name (if other than driver):_____

Address:_____   Phone Number:_____

Insurance Company for Other Vehicle:_____   Policy Number:_____

List Names of Any Passengers in Other Vehicle:_____

_____

List Names of Anyone Appearing or Claiming to Be Injured:_____

_____

Extent of Injuries:_____

Describe Damage to Other Vehicle:_____

USE BACK OF FORM TO DESCRIBE HOW ACCIDENT OCCURRED

Protect your driver's license: By law, as a driver of any vehicle involved in an accident, you **MUST** follow the following five steps. Suspension of your driver's license can result from failure to comply with these regulations.

IF YOU ARE INVOLVED AS A DRIVER IN AN ACCIDENT, YOU MUST:

1. Stop immediately.
2. Assist the injured if aid is needed or requested.
3. Exchange information with other persons involved.
4. Notify the local police, sheriff, or state patrol immediately.
5. Report the accident to the State Motor Vehicle Division and/or local police.

NAME	**REPORT OF THEFT, VANDALISM, OR DAMAGE TO EQUIPMENT, TOOLS, OR PROPERTY**
NUMBER	**JM-9**
ORIGINATOR	**Any employee**
COPIES	**1**
DISTRIBUTION	**Immediate supervisor**
SIZE	**8½ × 11 in.**
PURPOSE	On this document may be reported the theft, vandalism, or damage to equipment, material, tools, or property of the company.

The individual making out this report must also report the incident to the local police, and make a note of the details of the police report on this form.

REPORT OF THEFT, VANDALISM OR DAMAGE TO EQUIPMENT, TOOLS, OR PROPERTY

Report Date: _____ Date of Loss: _____ Time of Loss: _____

Name of Person who Discovered Loss: _____

Name of Supervisor: _____ Company Job Number: _____

Location of Loss: _____

(Street Address)   (City/County)   (State)

Type of Loss (Circle one.)   THEFT   VANDALISM   FIRE   OTHER (Explain) _____

Complete List of Stolen or Damaged Property and Approximate Value. (Use back of form if needed.)

_____

_____

_____

_____

_____

_____

_____

_____

_____

_____

_____

_____

_____

_____

_____

_____

_____

_____

_____

_____

_____

_____

Date and Time Reported to Police: _____

Police Department Reported to: _____

Case Number: _____ Investigating Officer: _____

NAME	**CONFIDENTIAL FOREMAN RATING**
NUMBER	**JM-10**
ORIGINATOR	**Construction Manager**
COPIES	**2**
DISTRIBUTION	**Construction Manager, Contract Manager**
SIZE	**8½ × 11 in.**

PURPOSE

Choosing the best person to run a job deserves much consideration, since this choice has a great influence on the success or failure of the job. The selection should never be made by any one person. Together, the Construction Manager and the Contract Managers should decide on the foreman for each job.

All estimates include foreman and general foreman rates, in strict accordance with the minimum requirements of the labor agreement. Therefore, any deviations from this criterion in making field assignments must be justified.

The worth of a foreman depends on how well he measures up to the requirements of a particular job. Not all good foreman are good on every type of job. There are square pegs and round pegs and square holes and round holes, and it is management's job to match the pegs with the holes.

Since the choice of a foreman should be based on personal experience, the prospective foreman's past record of accomplishment, general consensus, and overall judgment, it behooves the decision maker to collect opinions from qualified persons who have been in a position to observe the individual's capabilities and performance.

The Confidential Foreman Rating is used to rate foremen.

These rating sheets should be consulted as necessary in selecting foremen and should be revised as conditions change. Because of the confidential nature of such an evaluation, only one copy should be made, and it should be kept in the construction manager's file.

Each foreman is rated as follows:

A—excellent
B—good
C—fair
D—poor

In addition to the letter rating, supervisory ability also carries a number rating indicating the maximum number of workers that the foreman can effectively supervise.

The Confidential Foreman Rating provides space for rating each ability. The ratings are established on the basis of the collective judgment of the construction manager and the Contract Manager.

The Confidential Foreman Rating is based on the following factors:

Administrative Ability. Rating in this area indicates how well the foreman:

✓ Adapts to a system.
✓ Fits into the "team concept" of working together.
✓ Can receive and follow instructions.
✓ Adheres to standard procedures.
✓ Handles forms, reports, and paperwork in general.

Managerial Ability. Rating in this area indicates how well the foreman is organized, as evidenced by:

✓ The arrangement of material on the job.
✓ The arrangement of tools on the job.
✓ The arrangement of records, plans, and specifications.
✓ How he approaches the job—thinking things through and planning ahead.
✓ How he gives directions to the workers in the crew.
✓ How he manages the work and the workers.

Supervisory Ability. Rating in this area indicates the foreman's ability to supervise wireworkers effectively. This characteristic involves personality as well as principles. The measure of success is in the productivity of the workers supervised.

First, the foreman's personality and attitude must be such that wireworkers will want to get the work done. Second, he must have technical knowledge, or else workers will not want to work for him for very long. A foreman's ability to supervise workers may vary widely with the capabilities and personalities of the individual workers and the number of workers on the job.

Technical Ability. Rating in this area indicates the foreman's knowledge of:

✓ Basic electrical theory.
✓ Basic electrical engineering practices.
✓ Electrical equipment.
✓ Common electrical field practices.
✓ National and local electrical codes.
✓ General construction practices.

Potential Ability. Rating in this area indicates the foreman's potential ability to develop the four preceding abilities. This ability is largely a function of attitude and desire, provided, of course, that the basic physical and mental requirements are present.

Training is the final ingredient in developing one's full potential. A foreman who has the desire and the other basic abilities should be given every opportunity to receive training from all available sources. Foreman meetings provide an opportunity to determine those foremen who want to improve their abilities, as well as an opportunity to offer training in many areas.

CONFIDENTIAL FOREMAN RATING

Job Name: _C.F. & I. Ron Mill_   Date: _3_ / _15_ / _90_

| Name | Grade 1 to 10 (10 = best). | | | | | Remarks
Show remarks for each. |
	Administrative	Managerial	Supervisory	Technical	Potential	
J. Young	9	10	10	10	10	Rating A
R. Jones	5	5	9	10	9	Rating B
P. Pankey	5	5	8	9	8	Rating C

Foreman: _H. Gibbs_

CONFIDENTIAL FOREMAN RATING

Job Name: _____   Date: ____ / ____ / ____

Name	Grade 1 to 10 (10 = best).					Remarks Show remarks for each.
	Administrative	Managerial	Supervisory	Technical	Potential	

Foreman: _____

NAME	**REVISION TO BILL OF MATERIAL**
NUMBER	**JM-11**
ORIGINATOR	**Contract Manager**
COPIES	**3**
DISTRIBUTION	**Contract Manager, Purchasing, Job Foreman**
SIZE	**8½ × 11 in.**
PURPOSE	This document records changes in the Bill of Material.

This document records changes in the Bill of Material.

When the change involves the quantity and/or the description of an item, a Revision to Bill of Material form is used. If the change involves the date required, it is made on a Material Scheduling Change form. If the change involves the delivery date of only part of the total quantity, a new Bill of Material is written listing the quantity of the items and the new date required.

The main objective of this form is to advise the Purchasing Department of changes in the requirements of the job, as dictated by changes in the job plans or in the work made on the job, or when material is stolen.

The copies for the Contract Manager and the Job Manager are acknowledged and returned to them.

REVISION TO BILL OF MATERIAL

Job Name: _____   Date: _____

Work Order Number: _____   B/M Number: _____   P.O. Number: _____

Quantity	Description	Change From	To
_____	_____	_____	_____
_____	_____	_____	_____
_____	_____	_____	_____
_____	_____	_____	_____
_____	_____	_____	_____
_____	_____	_____	_____
_____	_____	_____	_____
_____	_____	_____	_____
_____	_____	_____	_____
_____	_____	_____	_____
_____	_____	_____	_____
_____	_____	_____	_____
_____	_____	_____	_____
_____	_____	_____	_____
_____	_____	_____	_____
_____	_____	_____	_____
_____	_____	_____	_____
_____	_____	_____	_____
_____	_____	_____	_____
_____	_____	_____	_____
_____	_____	_____	_____
_____	_____	_____	_____

Reason for Change:

_____   _____
Contract Manager   Purchasing Acknowledgment

NAME | **MATERIAL SCHEDULING CHANGE**

NUMBER | **JM-12**

ORIGINATOR | **Contract Manager**

COPIES | **3**

DISTRIBUTION | **Contract Manager, Purchasing Department, Job Foreman**

SIZE | **8½ × 11 in.**

PURPOSE | Record changes in material requirement dates on this form.

Material scheduling changes are usually the result of changing the Activity Schedule when the progress of the job is accelerated or falls behind the original schedule. This type of change is serious because it may affect a number of activities.

Nevertheless, the Activity Schedule should be reviewed frequently and appropriate changes made when required. The Activity Schedule (see Form JP-4), which is represented on the lower portion of this sheet, is a graphic representation of the best estimates of when each category of electrical work will occur for the entire job by dates of the month and by duration—the sequence events, their duration and the number of men required to perform the work. The tools required for the work are also included on the chart, and the time at which Foremen and General Foremen are required. The revision is accomplished by reworking the Activity Schedule Work Sheet and reconstructing the chart. It is very important to keep the Activity Schedule up-to-date because dates for purchasing and delivery of material to the job are based on it.

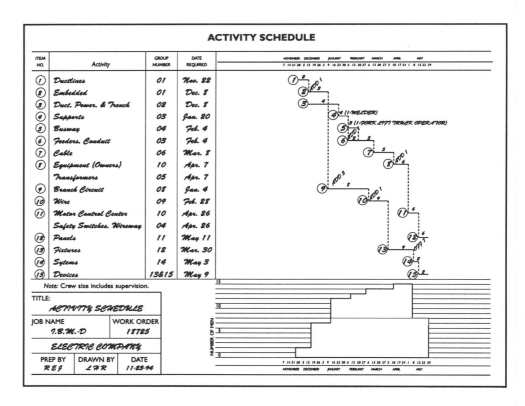

MATERIAL SCHEDULING CHANGE

Job Name: _____   Work Order Number: _____

Change 'Date Required' on Following Bills of Material as Indicated:

Group Number	Description	Purchase Order Number	Bill of Material Number	Date Required Changed	
				From	To

Signed: _____   Date: _____

Note: Purchasing is to notify the Contract Manager <u>immediately</u> if the above revised dates <u>will not</u> be met, stating reasons.

NAME
NUMBER
ORIGINATOR
COPIES
DISTRIBUTION
SIZE
PURPOSE

TOOL MAINTENANCE AND REPAIR

NUMBER JM-13

ORIGINATOR Field Foreman

COPIES 1

DISTRIBUTION Tool man (in warehouse)

SIZE 8½ × 11 in.

PURPOSE Jobs must be done with up-to-date tools of every kind. Equipping the job is only one step in the overall provision of tools. The others are:

✓ Purchase and rental.

✓ Storage.

✓ Maintenance and repair.

✓ Retirement.

✓ Records.

✓ Disbursement and return.

All these functions require special care and treatment, but the workers in the field are expected to take good care of all tools by doing the following:

✓ Using tools properly

✓ Performing proper field maintenance and storage

✓ Tagging all substandard tools and returning them to the shop for service

Regardless of how tools are used and cared for in the field, it is essential that all tools sent to the field be in excellent working condition, and that all parts mandatory for the operation of the tool accompany it to the field. When a tool is in need of service, the field man must fill out the Tool Maintenance and Repair form and return it to the shop with the tool.

TOOL MAINTENANCE AND REPAIR

Job Name: Littleton Library   Date: _____

	Tool parts Missing or Damaged	Dull Cutting Edges	Supplies Missing or Inadequate	Operation Electrical
Bag, screw			Replenish stock	
Bender, hydraulic	Shoes, pins, bars decals, etc.		1 pt of hyd. oil	Test pump motor
Box, cable-splicing			Replenish stock	
Box, gang	Hinges, hasp doors, rollers			
Can, oil-cutting or lube			Replenish oil	
Carts	Wheels			
Cord, extension	Caps and plugs			Test
Crimper, hydraulic (hypress)	Dies			
Drill, core	Adapters			Test
Drill, core sump pump				Test
Drill, core barnacle	Gasket			
Drills	Chuck key			
Drill, press	Chuck key			Test
Drill, press magnette	Chuck key			
Drive pump	Chuck			Test
Fastener, powder	Pins, etc.			
Generator			Gasoline	Test
Grinder, bench			Emery wheels	Test
Hammer, electric	Chisels, drills		Grease	Test
Hammer, drill phillips		Percussion bits	2½ in Percussion bits	Test
Hammer, drill (large-medium)		Drills		Test
Hose, garden	Connections		Gaskets	
Ladder	Steps, rungs sides			
Levels	Broken vials			
Puller, cable motorized	Cable			Test
Puller, cable manual	Cable			
Punch, KO hydraulic	Cutters, bolts	Cutters		
Punch, KO ratchet	Cutters, bolts	Cutters		
Saws, all kinds			3 extra blades	Test
Taps, fish vacuum	Parts			Test
Threaders	Dies	Dies		

Signed: _____

TOOL MAINTENANCE AND REPAIR

Job Name: _____

Tool parts Missing or Damaged	Dull Cutting Edges	Supplies Missing or Inadequate	Operation Electrical

Signed: _____

NAME	**SHOP DRAWING LOG**
NUMBER	**JM-14**
ORIGINATOR	**Contract Manager**
COPIES	**1**
DISTRIBUTION	**Contract Manager**
SIZE	**8½ × 11 in.**

PURPOSE

Most jobs require some electrical equipment of a special nature such as custom-made equipment for a special use. For such equipment, the manufacturer is required to furnish to the electrical contractor shop drawings for approval of the owner. A Shop Drawing Log provides a systematic method of recording the steps required to obtain this approval before proceeding with the manufacture, so that the equipment will be available and on the job at the proper time.

Generally, the owner, through the architect and/or the general contractor, requires this approval. However, when this approval is not required, it should be the policy of the company to require such drawings so that proper advance planning can be done in the field. On rare occasions, shop drawings are required for record purposes only. In any event, the procurement of shop drawings and their ultimate approval are becoming an increasingly important part of the electrical contractor's responsibility for getting the right material in the right quantity and at the right time.

Generally, these drawings require an approval before fabrication is authorized. In some cases, however, fabrication is authorized on the basis of the description on the Bill of Material, and shop drawings are required for record purposes only.

As far as the manufacturer is concerned, the only approval needed is that of the company. The approval of an architect or engineer alone is not sufficient to release equipment for manufacture.

As each step of this procedure is accomplished for any one transmittal, that information is shown on the Shop Drawing Log, together with the Shop Drawing Transmittal number and the transmittal number of the general contractor. In the event a resubmittal is required, the same procedure is repeated, using the same transmittal number plus a letter indicating the number of resubmittals (e.g., 3-A, 3-B, 3-C etc.).

Remember that it is the approval of the electrical contracting company that the manufacturer requires, not that of the architect or engineer. The company, not the architect, is the purchaser. It may require the architect's approval, and in most cases it does, but this is for the company's benefit and is not a sufficient basis on which to release the equipment for manufacture.

SHOP DRAWING LOG

Number of Copies Required:_____

(X+5)

Job Name: _____

Date:_____ / _____ / _____

Min 2 weeks

Max 2 weeks

Item	Manu-facturer	Bill of Material	P.O.	Date Ordered	Date Required	Date from Purchase	Date Preliminary to Field (A)	Date to General Contractor (B)	Date from General Contractor	Action A Approved A/C Approved as Corrected D/R Disapproved Resubmit	Date to Purchase (C)	Date Approved Copies to Field (D)

NAME	# SHOP DRAWING TRANSMITTAL FORM
NUMBER	**JM-15**
ORIGINATOR	**Contract Manager**
COPIES	**1**
DISTRIBUTION	**Contract Manager**
SIZE	**8½ × 11 in.**

PURPOSE

When special electrical equipment is required, the Shop Drawing Transmittal Form is used to transmit shop drawings to the field, to the General Contractor, and to the Purchasing Agent.

The number of copies of shop drawings required initially from the manufacturer should be determined by an inquiry to the general contractor as to his requirements, exclusive of the number of approved copies that will be returned to the company, plus the five copies the office will require (one preliminary copy for the field, one preliminary copy for the office, one copy approved for the field, one copy approved for the office, and one copy approved for Purchasing).

The request for the required number of copies ($x+5$) should be made on the Bill of Material, which also gives a description of each drawing, stating whether it is for approval or for record purposes, and the date it is required. When drawings are requested for approval, under no circumstances should material be released for shipment until approval of the applicable shop drawings is received by the manufacturer. Purchasing should expedite shop drawings in the same manner as it expedites the material.

When the required number of copies ($x+5$) are received from the manufacturer (through Purchasing), the Contract Manager should examine the drawings to see that they conform to the specifications and that they meet any additional requirements shown on the Bill of Material. When the Contract Manager is satisfied that the shop drawings do, in fact, meet all the requirements or when he has added notes or made corrections to all copies such that they will meet all the requirements, he sees that all copies ($x+5$) are stamped "Approved," dated, and initialed. At this time the drawings are stamped and dated to indicate the latest acceptable date for the return of approved drawings if the construction schedule is to be maintained.

One copy of these approved preliminary drawings should be transmitted to the field via the Shop Drawing Transmittal form. Another copy of these drawings is placed in the job file, together with a copy of the field transmittal.

All remaining copies ($x+3$) should then be transmitted to the general contractor via the Shop Drawing Transmittal form, and a copy of the transmittal form should be placed in the job file. At this point, the Contract Manager should place a note in his tickler file (or the equivalent) to remind him to expedite drawings from the general contractor if they have not been received by the latest acceptable date stamped on each drawing. This expediting function is strictly the Contract Manager's responsibility, not Purchasing's. In the event the Contract Manager is not sucessful in obtaining the approved drawings at least in one additional week (sooner is necessary), the matter should be turned over to the construction manager.

SHOP DRAWING TRANSMITTAL FORM

To: _____   Date: _____

_____   Bill of Material Number: _____

_____   P.O. Number: _____

Attention: _____   Electrical Contractor Transmittal Number: _____

Job Name: _____   General Contractor Transmittal Number: _____

Attached herewith are the following shop drawings as prepared by

_____   _____

_____   _____

which have been

☐ Prepared for your approval   ☐ Approved   ☐ Disapproved, resubmit

☐ Corrected and/or revised   ☐ Approved as   ☐ Prepared for your records
   for your approval   noted   These are FINAL

Number of Copies	Drawing Number	Description	Date of Drawing

☐ Release for shipment as scheduled.

☐ Please return three (3) "APPROVED" copies.

_____

NAME	**FEEDER CUTTING LENGTHS**
NUMBER	**JM-16**
ORIGINATOR	**Contract Manager**
COPIES	**2**
DISTRIBUTION	**Contract Manager, Purchasing Department**
SIZE	**8½ × 11 in.**

PURPOSE On this form, the Contract Manager records for Purchasing the feeder cables from the estimator's takeoff, which are listed by the number of conductors on each reel, as well as by their size, type, and color. The estimator takes off the feeder runs with the confirmation of the processor, and the required footage of cable is shown on the Bill of Material in total footage. It is hazardous to cut the cable and reel it for usage on the job without field-measuring it beforehand.

Another very important consideration when purchasing the cable for the job is the estimated price in the contract. Sometimes a whole year expires between the time of the estimate quote and the time at which the cable is to be pulled. In a volatile copper or aluminum market, the price of wire can change appreciably. To protect the estimate price, a commitment must be made to purchase the cable at that price, with the understanding that the cable will be cut and reeled according to field measure. Therefore, a precise cable-handling procedure must be followed in the operation, as follows:

FEEDER CUTTING LENGTHS

Job Name: _Empire State Building_   Work Order Number: _18971_
Date Requested: _____   Date Originally Scheduled: _____

Feeder Number	Number of Pieces	Size	Insulation Type	Insulation Color	Conductor Material	Length	Packaged	B/M	P.O.	Date Required
LPA	3	3/0	THW	Blk.	Al.	98 ft	3 reels	18972	39056	10/31
PP	3	3/0	THW	Blk.	Al.	90 ft	3 reels	18972	39056	10/31
PP	1	3/0	THW	Blk.	Al.	87 ft	1 reel	18972	39056	10/31
Service	3	250	THW	Blk.	Al.	132 ft	3 reels	18972	39056	10/31
Service		250	THW	Blk.	Al.	128 ft	1 reel	18972	39056	10/31
Elevator	3	2	THW	Blk.	Cu.	73 ft	3 coils	18972	39056	10/31
Elevator	1	2	THW	Wht.	Cu.	78 ft	1 coil	18972	39056	10/31

✓ A Bill of Material is prepared by the Contract Manager, showing a feeder listing similar to that shown on the Feeder Cutting Lengths form; in the Bill of Material, each feeder is identified by plan number and described in the Where Used column.

✓ Requirement dates are established for both the regular Bill of Material and the feeder listing.

✓ Orders are placed by Purchasing according to the Bill of Material, with the understanding that quantities will depend on field measure at the time of requirement.

✓ When the exact cutting lengths have been determined by the field, they are transmitted to the Purchasing Agent through the Contract Manager on the standard Bill of Material.

FEEDER CUTTING LENGTHS

Job Name: _____ Work Order Number: _____

Date Requested: _____ Date Originally Scheduled: _____

Feeder Number	Number of Pieces	Size	Insulation		Conductor Material	Length	Packaged	B/M	P.O.	Date Required
			Type	Color						

| | | | NAME | **BILL OF MATERIAL FOR CABLE FIELD MEASURE** |

NAME **BILL OF MATERIAL FOR CABLE FIELD MEASURE**

NUMBER **JM-17**

ORIGINATOR **Contract Manager**

COPIES **3**

DISTRIBUTION **Purchasing, Contract Manager, Field Foreman**

SIZE **8½ × 11 in.**

PURPOSE On this form, the Contract Manager can finalize the field measured cable lengths according to:

1. The number of conductors.
2. Cable size.
3. Type of conductor.
4. Color of insulation
5. Location of run.

With this information, the Purchasing Agent can release the order from the supplier.

BILL OF MATERIAL FOR CABLE FIELD MEASURE

Floor Number: _____ Group: _(Cable field measure)_ Lead Time: _____ Sheet Number: _6a_

Job Name: ___Empire State Building___ Ship to: __Job-site__ Date: ___/___/___

Contr. Number: _____ Engineer: _____ Contract Manager: _V.7._ Work Order Number: _18971_

Quantity Shipped	Quantity Ordered	C L	Item Code	Description		Where Used	Date Required	P.O Number
	3 pcs.		330423	Wire. Al. 3/0. THW Black	98 ft	LPA	10-31	
				(3-reels)				
	3 pcs.		330423	Wire. Al. 3/0. THW Black	90 ft	PP		39056
	1 pc.		330423	Wire. Al. 3/0 THW Black	87 ft	PP		
				(4-reels)				
	3 pcs.		330425	Wire. Al.250 MCM. THW Black	132 ft	Service		
	1 pc.		330425	Wire. Al.250 MCM. THW Black	128 ft	Service		
				(4-reels)				
	3 pcs.		331402	Wire. Cu.#2. THW. Black	73 ft	Elevator		
	1 pc.		331402	Wire. Cu. #2. THW White	78 ft	Elevator		
				(4-coils)				

BILL OF MATERIAL FOR CABLE FIELD MEASURE

Floor
Number: _____ Group: _____ Lead Time: _____ Sheet Number: _____

Job Name: _____ Ship to: _____ Date: ____ / ____ / ____

Contr. Number: _____ Engineer: _____ Contract Manager: _____ Work Order Number: _____

Quantity Shipped	Quantity Ordered	C L	Item Code	Description	Where Used	Date Required	P.O Number

NAME	**JOB CLOSEOUT CHECKLIST**
NUMBER	**JM-18**
ORIGINATOR	**Contract Manager**
COPIES	**3**
DISTRIBUTION	**Contract Manager, Construction Manager, General Manager**
SIZE	**11 × 8½ in.**

PURPOSE   The importance of closing out the job properly cannot be overemphasized. Three principal reports must be completed as the job comes to a close:

1. Job Closeout Checklist
2. Job Evaluation Report
3. Estimate Evaluation Report

It is very important that all jobs be closed out as quickly as possible after the work has been completed. Failure to do so results in a delay in collecting the final payment and the retainage, which usually amounts to 10 or 15% of the total contract price.

On most construction jobs, many man-hours are wasted during the final stages of construction because of failure to account for all minor items of material required to complete the job. End-of-job waste of man-hours may be caused by the following specific problems:

✓ Material shortages
✓ The need to replace fixture glass, box covers, panel trims, switch and receptacle plates, etc.
✓ Speciality items like systems
✓ Damaged lighting fixtures
✓ Final connections to late arriving equipment
✓ Punch list items and inspection reports
✓ Panel directories
✓ Final decisions by the architect or owner on questionable items
✓ Waiting for other crafts to clear conflicts
✓ "Fine print" items on plans and in specifications

For most jobs, there is very little allowance made on the Bid Summary Sheet for such problems. They should therefore be dealt with as the work progresses and not allowed to drag out to the last minute as special items.

The Job Closeout Checklist is especially important to fill out well in advance of the completion date because it can be time-consuming to complete.

JOB CLOSEOUT CHECKLIST

Job Name: _____ Work Order Number: _____

Contract Manager: _____ Job Foreman: _____ Date: __ / __ / __

Field Work	Date Completed
1. Conduct Prefinal Inspection to be made by Contract Manager with Job Foreman prior to final inspection by the Architect or City, County, or State Inspector. Check for the following items:	
a. General compliance with plans and specifications	
b. Covers, plates, panel trims, plastic or glass parts DAMAGED OR MISSING.	
c. Panel directories (typed and in place)	
d. Trash removal and General cleanup	
e. Company name plates on panels, switchgear, etc.	
2. Call for final inspection by City, County or State Inspector. Make certain inspector signs the permit card on the job.	
3. Complete all punch list or deficiency items.	
4. Ready tools, material, trailer and/or equipment for pickup. Be sure yellow tags are on all defective or inoperative tools.	
5. Indoctrinate operating personnel as required by the specifications.	
6. Deliver the following items as required by the specifications.	
a. Spare fuses	
b. Keys to panels	
c. Spare parts	
GET SIGNED RECEIPT FOR ABOVE ITEMS	
Office Work	
1. Write memo to Sales Manager requesting a Salesman be assigned this customer and recommend a time to visit the job with the Salesman to introduce him to the proper customer's Representative. This should be done at least one week before our men are off the job.	
2. At least one week before men will be off the job, obtain a complete list of tools still charged to the job from the tool man and give to the Job Foreman for an accounting. All tools must be accounted for in one way or another.	

	Date Completed
3. Transmit the following items as required by the specifications:	
a. "As Built" drawings	
b. Operation and maintenance manuals	
c. Cost data, spare parts list, etc.	
4. Write letter as required by the specifications, confirming the guarantee requirements (use the same wording as in the specifications). Cite the date acceptance or beneficial occupancy from which the year's guarantee is to be determined. If a date has not been formally established—establish one.	
5. Ascertain that all field copies of all work orders and change orders are completed, signed, dated, and transmitted to accounting.	
6. Ascertain that all required sales tax and payroll documents have been issued by accounting.	
7. Ascertain that all FINAL billings have been rendered.	
8. Remove job from the "Active Job List."	
9. Issue Close-Out Ticket to Warehouse	
10. Wire a narrative explaining the outcome of the job. Transmit to Construction Manager. Include Job Close-Out Check Sheet, and Estimate Evaluation Sheets.	
11. After FINAL payment is received and ALL job responsibilities have been completed, DESTROY all papers in the Job File except the following:	
a. One complete set of latest plans	
b. One complete set of specifications and addendas	
c. All correspondence (both incoming and outgoing)	
d. One set of approved shop drawings	
e. Any other documents deemed worth saving Transmit these items to the draftsman for proper storage.	
12. Write letter of appreciation to Foreman, if warranted.	

Remarks: _____

_____

_____ Signed _____

NAME	**JOB EVALUATION REPORT**
NUMBER	**JM-19**
ORIGINATOR	**Construction Manager**
COPIES	**3**
DISTRIBUTION	**General Manager, Accounting, Construction Manager**
SIZE	**8½ × 11 in.**
PURPOSE	The report on the results of the job is a written report prepared by the Contract Manager; it is intended to give an overall impression of all important aspects of the job. The Job Evaluation Report is prepared by the Construction Manager after receiving the written report on the results of the job from the Contract Manager.

The Job Evaluation Report is the result of accounting for differences between the estimated amounts and the actual dollars spent. To do this satisfactorily, data has to be accumulated during the entire course of the job, but in general these differences occur because of the following:

✓ Errors in the estimate as a result of:
    Faulty arithmetic
    Incorrect quantities
    Incorrect types of material
    Incorrect labor unit application

✓ Failure to purchase according to estimate

✓ Failure to deliver material and equipment as scheduled

✓ Failure to manage the job properly because of:
    Inefficient manpower
    Inadequate or incorrect information
    Inefficient methods and tooling
    Poor and inadequate planning and scheduling
    Inadequate and ineffective supervision

✓ Circumstances beyond the contractor's control
    Weather conditions
    The general contractor's performance
    Theft of materials

When an error is discovered in an estimate, it should be reported immediately in an effort to assist in improving estimating accuracy, as well as to aid in accounting for any overall differences between estimated amounts and the actual job cost.

JOB EVALUATION REPORT

Job Name: _____ Work Order Number: _____

Salesperson: _____ Estimator: _____ Processor: _____

Contract Manager: _____ Foreman: _____ General Contractor: _____

	Estimated		Actual	Percent of Final Estimated	Percent of Change
	Original	Final			
Material	$	$	$		
Labor	$	$	$		
Job Expense	$	$	$		
Prime Cost	$	$	$		
Margin in $	$	$	$		
Margin in %	%	%			
Total Contract	$	$	_____	_____	

CONTRACT MANAGER'S EXPLANATION OF RESULTS: Use facts, figures, opinions.
Attach accumulated Estimate Evaluation Reports, Group Labor Reports

Date: _____ Signed: _____
(Use additional sheets if necessary.)

NAME	**ESTIMATE EVALUATION REPORT**
NUMBER	**JM-20**
ORIGINATOR	**Construction Manager or Contract Manager**
COPIES	**3**
DISTRIBUTION	**General Manager, Estimator, Construction Manager**
SIZE	**8½ × 11 in.**

PURPOSE | The Estimate Evaluation Report is used at any time during the job to account for errors in various forms of estimating that should be corrected. In addition to the self-explanatory items on the form, it must show:

✓ The percentage complete of the job.
✓ A complete description of the error, and how it should have been handled.
✓ The approximated costs of material and labor.
✓ Whether the error is a loss, an omission, or an excess with a positive connotation.

Note: An excess is an error as much as a loss is, and it is important to recognize it as such.

Completing a job and closing it out expeditiously may save it from being a loser, and can prevent many disagreeable moments for the owner, architect, general contractor, and above all the electrical contractor. Job closeout must therefore be done in a methodical manner to ensure that it will be an effective operation.

ESTIMATE EVALUATION REPORT

Report Number: _____   Job _____ %

Date: ____ / ____ / ____

Job Name: _____   Work Order Number: _____

Estimator: _____   Salesperson: _____   Processor: _____   Contract Manager: _____

Number	Description	By	Material $	Labor $	Total $	+ or −

NAME	**CHANGE ORDER**
NUMBER	**JM-21**
ORIGINATOR	**Contract Manager**
COPIES	**4**
DISTRIBUTION	**Accounting, field, Construction Manager, Contract Manager**
SIZE	**8½ × 11 in.**
PURPOSE	On this form, record all the details relative to a Change Order.

Change Orders are of two general types: Contract Change Orders and Time and Material Change Orders.

Regardless of the type, Change Orders are not particularly desirable, but they do occur on almost every job, and it is very important that they be handled properly. The information required is as follows:

✓ *Authorization.* The appropriate box is checked, and the Purchase Order number is shown, if available. If authorization is verbal, the individuals involved are identified, and the date of the authorization is given, as well as the date of the confirming letter sent by the Contract Manager.

✓ *Work to Be Done.* The appropriate box is checked, and sufficient instructions are given so that the person on the job can accomplish the work.

✓ *Bills of Material.* The appropriate box is checked.

✓ *Architectural Plans.* The appropriate box is checked.

✓ *Company Drawing.* The appropriate box is checked.

✓ *Billing.* The appropriate box is checked.

✓ *Signed.* The signature of the Contract Manager.

✓ *Billing Information.* This includes the date for invoicing, the number of invoices, the percent retainage, the number of payroll reports, the number of payroll affidavits, and the number of sales tax affidavits. These items are almost always the same as in the base contract, in which case "same as base" is indicated.

Under no circumstances should any extra work be performed in the field without a Change Order. When the work must proceed before an agreement has been reached as to who will ultimately pay for it, a Change Order is written and issued to the field immediately, before the work is started; the liability is determined later.

Time and Material Change Orders are seldom authorized in writing, and therefore confirming letters are almost always required. In addition to confirming the fact that orders to proceed were issued, it is also important to identify the person authorizing the work and to define the scope of the work.

Time and Material Change Orders: Customers want work done on a time and material basis for basically three reasons:

1. There is not enough time to obtain a proposal.
2. The work is not defined well enough to obtain a proposal.
3. They believe they can accomplish the work more economically.

Regardless of the reason, it must be pointed out to the field forces that they are totally responsible for the accuracy of all time and material billings. Unless every minute used in the accomplishment of the work is listed on the Time Report, it will not be paid for. This time must include a portion of the General Foreman's and the Foreman's time, all the time required to maintain records and complete forms, and all the time required to actually do the work. Time and Material Change Orders are usually less desirable than Contract Change Orders. In most cases, if there is a choice, the Contract Change Order should be requested.

Contract Change Orders: When a customer requests a quotation on a change he wishes to make in the electrical work, the Contract Manager must decide how the estimate is to be prepared. There are three possibilities:

1. If the change is well defined on a drawing and/or in some sort of a specification and if the estimating time would be 2 or 3 hours or more, he should probably ask the Estimating Department to prepare the estimate.
2. If a rather extensive change is described to the Contract Manager over the telephone and if there are no drawings, he may ask the processor, the draftsman, or the construction secretary to assist him in preparing the estimate.
3. If the change is small, he can prepare the estimate himself.

It is not uncommon for Change Orders to include changes to items of material that are already on the job or at least are ordered especially for that job. In such cases, it is necessary to include in the estimate any restocking charge that may be involved. This amount will appear as a reduction to the credit to be offered for the unused item. The best thing to do, if possible, is to suggest to the customer that he keep the item for future use and offer no credit at all.

If a credit is requested, however, the amount of the restocking charge that the manufacturer will impose should be obtained from Purchasing. If the item is already on the job, the contractor can impose an additional restocking charge because of his extra handling expense. Returning items for credit should be avoided if possible, but when it is necessary, restocking charges must be assessed without compunction and with full appreciation for the impact it has in all areas.

When the customer accepts the proposal, a Change Order number is assigned, and the appropriate entry is made in the Change Order record.

CHANGE ORDER

Date: _5_ / _1_ / _XX_

Number: _1A_

Promised to Start

Date: _5_ / _1_ / _XX_   Time: _8:00_   Associated Work Order Number: _18276_

Job Name: _Littleton Library_

Job Location: _4220 Broadway_

Charge to: _Mead + Mount Const. Co._

Address for Invoice: _218 Denver Club Bldg._

AUTHORIZATION ☐ Contract   ☐ Letter   ☐ Purchase Order   Number: _1A_

☐ Verbal order (Confirming letter dated _4-21-XX_)

From: _J. Jackson_   To: _Earl Underwood_   Date: _4_ / _2_ / _XX_

WORK TO BE DONE as per ☐ Plans ☐ Specifications   Alternative Number: _____   Addendum Number: _____

☐ Drawing number: _None_   ☐ Following Description: _____

Install wiring for air conditioning in main area.

_____

_____

_____

_____

_____

Bills of Material	☐ In Purchasing		☐ To Follow	☒ Not Required
Architectural Plans	☐ Attached	☐ In Field	☒ To Follow	☐ Not Available
Sturgeon Drawing	☐ Attached		☒ To Follow	☐ Not Required
Billing	☐ Contract	☐ T&M w/guaranteed maximum	☒ T&M ☐ Unit Prices	☐ Cost Plus

Signed: _____

BILLING INFORMATION

Date for Invoicing: _6-1-XX_   Number of Invoices: _3_   Percent Retainage: _None_

Number of Payroll Reports: _____   Number of Payroll Affidavits: _____   Number of Sales Tax Affidavits: _____

☐ Contract $ _____   Material $ _____   Margin $ _____

_____   Labor $ _____   Markup   _____ %

Total for   Job Expense $ _____   Man-hours   _____

Comb. Job $ _____   Prime Cost $ _____

☒ T&M   Labor Rate: _x_   per hr. or Labor Cost + Taxes + Burden + _15% + 10%_

Materials: _15+10_   Equipment   %_Total Not to Exceed $ _____

☐ Cost Plus a Fixed Fee   ☐ Cost Plus   ☐ Unit Prices   ☐ Formula   ☒ Other

Explain: _Materials to be billed end price—20%_

CHANGE ORDER

Date:____ / ____ / ____

Number: _____

Promised to Start

Date:____ / ____ / ____   Time:_____   Associated Work Order Number: _____

Job Name:_____

Job Location: _____

Charge to: _____

Address for Invoice: _____

AUTHORIZATION ☐ Contract   ☐ Letter   ☐ Purchase Order   Number: _____

☐ Verbal order (Confirming letter dated _____)

From: _____   To: _____   Date:____ / ____ / ____

WORK TO BE DONE as per ☐ Plans   ☐ Specifications   Alternative Number: _____   Addendum Number: _____

☐ Drawing number: _____   ☐ Following Description: _____

_____

_____

_____

_____

_____

Bills of Material	☐ In Purchasing		☐ To Follow	☐ Not Required
Architectural Plans	☐ Attached	☐ In Field	☐ To Follow	☐ Not Available
Sturgeon Drawing	☐ Attached		☐ To Follow	☐ Not Required
Billing	☐ Contract	☐ T&M w/guaranteed maximum	☐ T&M ☐ Unit Prices	☐ Cost Plus

Signed: _____

BILLING INFORMATION

Date for Invoicing: _____   Number of Invoices: _____   Percent Retainage: _____

Number of Payroll Reports: _____   Number of Payroll Affidavits: _____   Number of Sales Tax Affidavits: _____

☐ Contract	$ _____	Material	$ _____	Margin	$ _____	
	_____	Labor	$ _____	Markup	_____ %	
Total for		Job Expense	$ _____	Man-hours	_____	
Comb. Job	$ _____	Prime Cost	$ _____			

☐ T&M   Labor Rate: _____   per hr. or Labor Cost + Taxes + _____ %

Materials: _____   Equipment____ % Total Not to Exceed $ _____

☐ Cost Plus a Fixed Fee   ☐ Cost Plus   ☐ Unit Prices   ☐ Formula   ☐ Other

Explain: _____

NAME	**CHANGE ORDER BID SUMMARY**
NUMBER	**JM-22**
ORIGINATOR	**Contract Manager**
COPIES	**3**
DISTRIBUTION	**Construction Manager, Contract Manager, Accounting**
SIZE	**8½ × 11 in.**

PURPOSE   The Contract Manager uses this form to price out a Change Order according to the instructions on it.

The Bid Summary Sheet is used for T&M Orders or Field Orders, which are priced out according to the actual billing of material items, labor hours used, and equipment used.

It must be realized that industry allowances formed by previous experience predominates over actual business costs. Small contractors cannot survive on the 15% and 10% margins that are allowed.

A separate sheet of paper listing the items of material should accompany the Bid Summary Sheet.

CHANGE ORDER BID SUMMARY

Job Name: _____ Change Order Number: _____

Description of Change: _____

_____

MATERIAL COSTS $ _____

Labor Hours   _____

    Supervision: _____ %

    Job Factor: _____ % of _____   _____

      Total Man-hours   _____

<u>TOTAL LABOR COSTS</u> _____ hrs. @ _____ /hr. _____ $ _____

Job Expense

    Payroll Tax: _____ % of labor costs $ _____

    Rental of tools, equipment, etc.

    _____ $ _____

    Sales Tax: _____ % of $ _____ $ _____

    Engineering: _____ hours _____ $ _____

    Other Job Expense: _____

    _____ $ _____

TOTAL DIRECT JOB EXPENSE $ _____

    Total Prime Cost $ _____

    Overhead: _____ % $ _____

    Total Gross Cost $ _____

    Profit: _____ % $ _____

                                           Total Cost $ _____

    Bond: _____ % $ _____

TOTAL SELLING PRICE $ _____

Prepared by: _____ Checked by: _____

CHAPTER EIGHT
JOB PREPARATION FOR FIELD

FORM NO.	FORM NAME	PAGE
JP-1	Job Processing Check Sheet	232
JP-2	Proposed Material List	234
JP-3	Activity Schedule Work Sheet	236
JP-4	Activity Schedule	238
JP-5	Bill of Material	242
JP-6	Job Expense Requisition	244
JP-7	Fixture Schedule	246
JP-8	Panelboard Schedule	248
JP-9	Work Order	250

NAME	**JOB PROCESSING CHECK SHEET**
NUMBER	**JP-1**
ORIGINATOR	**Job Processor**
COPIES	**1**
DISTRIBUTION	**Job Processor**
SIZE	**8½ × 11 in.**

PURPOSE   The work required to get a job started for the field is called *job processing*. It is normally performed by the Job Processor under the direction of the Contract Manager in the Job Management Department. It is essential for effective operation of the business to prepare the work for the field. There are many important requirements of the job over which the field workers have no control, but which have significant bearing on the job. So it is important not only to provide the answers to these questions but also to support the field on a continuing basis for smooth operations.

While there are numerous details that must be performed by the processor, the main responsibilities are the following:

✓ Preparing plans and specifications for field use
✓ Preparing the Activity Schedule for material and manpower scheduling
✓ Preparing a Bill of Material with processor modifications
✓ Handling manufacturers' shop drawings
✓ Preparing tool list and requirement dates
✓ Preparing fixture schedule
✓ Preparing panelboard schedule

Most of the information required on the job is conveyed to the field by plans and specifications, but additional information is nearly always required to provide accurate, complete, and comprehensive information. The information that is usually required to supplement the contract drawings is the following:

✓ Wiring diagrams
✓ One-line diagrams
✓ Panelboard schedules
✓ Fixture schedules
✓ Equipment room layout
✓ Feeder conduit layout
✓ Busway routing
✓ Underfloor duct dimensional layout

Information such as this ensures adequate physical space, determines actual material requirements, and greatly facilitates installation procedures.

This form provides guidelines for the Job Processor to cover all the required items in the preparation of the paperwork for the job in the field.

JOB PROCESSING CHECK SHEET

Job Name: _____   Date:____ / ____ / ____

No.	Description	Status	No.	Description	Status
1	Check job file transfer sheet.		19	Prepare estimate evaluation.	
2	Assign work order number.		20	Prepare cards for manpower and contract manager's desk.	
3	Enter on active job list.		21	Check with utility on service.	
4	Enter in job log.		22	Check with telephone company on service.	
5	Assign contract manager.		23	Check with city and county for plan approval.	
6	Issue item tab to purchasing with wire comments.		24	Prepare shop drawing log.	
7	Prepare job file.		25	Prepare material status record.	
8	Apply for permit.		26	Order computerized bill of material and drawings.	
9	Write work order and distribute immediately.		27	Prepare change order records.	
10	Send group breakdowns to accounting.		28	Check on introduction letter.	
11	Bind specs in loose-leaf book. For field and Contract Manager indicate all addendas. Glue in specs and on plans.		29	Review job with estimator.	
			30	Collaborate with purchasing department on proposed material list and send to general. contractor.	
12	Place office copy of plans on stick and mark.		31	Prepare activity schedule.	
13	Edge prints for field.		32	Correct computerized bill of materials.	
14	Contract—original to accounting; copy to file.		33	Prepare manual bills of material, include fixture cuts and shop drawings.	
15	Send insurance certificates.		34	Prepare tool list.	
16	Request construction schedule from general contractor by letter.		35	List prefab possibilities.	
17	Request sales tax exemption.		36	Prepare job record box.	
18	Notify secretary as to file and action copy required.		37	Prepare detailed drawing for field.	

NAME	**PROPOSED MATERIAL LIST**
NUMBER	**JP-2**
ORIGINATOR	**Job Processor**
COPIES	**3**
DISTRIBUTION	**Purchasing, Contract Manager, Job Processor**
SIZE	**8½ × 11 in.**

PURPOSE — This is a list of the materials and equipment to present to the architect for approval to be used on the job. The list is submitted through the General Contractor.

In many cases the specifications provide a reference to the type of materials to be used on the job, but it is also possible to substitute materials of an equal type if approved.

The Proposed Material List is also used by the Purchasing Agent to furnish the Job Processor with the names of manufacturers who can furnish materials that meet the specifications.

The electrical contractor's most important function is to furnish the job with the right material, in the right quantity, at the right time. This is very difficult. To do so requires a systematic procedure, followed in detail and understood by everyone involved.

PROPOSED MATERIAL LIST

Job Name: _____ Work Order Number: _____

Architect: _____ Electrical Engineer: _____

Date: ____ / ____ / ____ Prepared by: _____ Approved by: _____

Estimator: _____ Job Process Engineer: _____

Items	Specified	Proposed	Enclosed Literature

NAME **ACTIVITY SCHEDULE WORK SHEET**

NUMBER **JP-3**

ORIGINATOR **Job Processor**

COPIES **2**

DISTRIBUTION **Contract Manager, Job Processor**

SIZE **8½ × 11 in.**

PURPOSE The Activity Schedule (Form JP-4) is the basis for all job scheduling. An *activity* is defined as a stage of construction that must be distinguished from other stages of construction so that the unavailability of an item of material or a piece of equipment does not interrupt the continuity of the work.

The value of material for an activity is not considered in the preparation of the Activity Schedule; only labor man-hours are considered for each activity. The estimator must therefore take into consideration that, to permit the Job Processor to prepare this schedule, the estimate must be subdivided by man-hours according to the work schedule. The Activity Schedule must also conform to the General Contractor's construction schedule.

The preparation of the Activity Schedule begins with the Activity Schedule Work Sheet. Each work schedule represents an activity, but certain items of concern within the work schedules, which usually involve embedded materials, are also activities and must be represented by an estimate of man-hours. These items may be the following:

ACTIVITY SCHEDULE WORK SHEET

Job Name: TBM—Package D   Date: Nov. 18, 19XX

No.	Activity Description	Man-hours	Material Group	Crew Size	Crew Weeks	Starting Date	Finishing Date
1	Ductlines	186	01	2	2.32	Nov 22	Dec 8
2	Embedded	462	01	3	3.85	Dec 8	Jan 4
3	Power and trench duct	965	02	4	6.03	Dec 8	Jan 20
4	Supports	360	03	4	2.25	Jan 20	Feb 4
5	Busway	252	04	3	2.10	Feb 4	Feb 21
6	Feeders	168	03	2	2.10	Feb 4	
		473		5	2.36		
7	Cable	883	06	5	4.41	Mar 8	Apr 7
8	Equipment (camera) transformers	619	10 05	6	2.58	Apr 7	Apr 26
9	Branch circuit	2469	08	8	7.72	Jan 4	Feb 28
10	Wire	1576	09	9	4.38	Feb 28	Mar 30
11	Motor control center safety switches, wireway	516	10 04	6	2.15	Apr 26	May 11
12	Panels	418	11	6	1.74	May 11	May 23
13	Fixtures	1730	12	9	4.80	Mar 30	May 3
14	Systems	66	14	2	.82	May 3	May 4
15	Devices	154	13 & 15	2	1.92	May 9	May 23
Total		11,297					

✓ Anchor bolts
✓ Underfloor duct
✓ Manhole fittings
✓ Fixture plaster frames
✓ Hanger rods
✓ Hanger racks
✓ Switchboard supports
✓ Panelboard tubs
✓ Templates
✓ Sleeves
✓ Pipe clamps

The first step in the preparation of the Activity Schedule is the completed Work Sheet. Nearly all the information except the chart is contained in this document, and, as stated above, the estimate must be structured to furnish this information as required for the Work Sheet.

It must be emphasized that each activity represents a division of the estimate, in this case 15 different man-hour subdivisions of the estimate. Preparing a manpower schedule for the job requires accurate man-hour subdivisions of the estimate.

ACTIVITY SCHEDULE WORK SHEET

Job Name: _____   Date: _____

No.	Activity Description	Man-hours	Material Group	Crew Size	Crew Weeks	Starting Date	Finishing Date

Total

NAME	**ACTIVITY SCHEDULE**
NUMBER	**JP-4**
ORIGINATOR	**Job Processor**
COPIES	**5**
DISTRIBUTION	**Contract Manager, Purchasing Department, General Manager, Construction Manager, Field Foreman**
SIZE	**8½ × 11 in.**

PURPOSE The Activity Schedule is the most important job management device available to supervision and top management. Yet its accuracy is subject to the continuity of the construction management of the general contractor, and it must be changed if the schedule for the job changes because the electrical contractor's material requirement dates are predicated on it.

In addition to the material requirement dates, the schedule may be used to:

✓ Determine manpower requirement dates.
✓ Determine tool requirement dates.
✓ Determine the activities involved in each subdivision of work when there is more than one subdivision.
✓ Provide a means of determining in advance the crew size.
✓ Provide a means of measuring crew accomplishment on site.

After the Activity Schedule Work Sheet is completed, the information is transferred to the Activity Schedule, which is prepared on ¹/₁₀-in. graph paper in the manner illustrated. The activities are listed on the left-hand side of the sheet in a manner that allows the graph line to be projected off them. A notation is made on the graph line as to the size of the crew, and dotted lines are used to show the sequence of the crew activities. In addition, a manpower chart is created by projecting the crew size from the graph lines down to the manpower chart below.

Since the value of material for an activity is to considered in the preparation of the Activity Schedule (only labor man-hours are considered), the estimator must take into consideration that, for the Processor to prepare the schedule, the estimate must be subdivided accordingly. The Activity Schedule must also conform to the general contractor's construction schedule.

The preparation of the Activity Schedule begins with the Activity Schedule Work Sheet. Each work schedule represents an activity, but certain items of concern within the work schedules, which usually involve embedded materials, are also activities and must be represented by an estimate of man-hours. These items may be the following:

- ✓ Anchor bolts
- ✓ Underfloor duct
- ✓ Manhole fittings
- ✓ Fixture plaster frames
- ✓ Hanger rods
- ✓ Hanger racks
- ✓ Switchboard supports
- ✓ Panelboard tubs
- ✓ Templates
- ✓ Sleeves
- ✓ Pipe clamps

For the most part, the progress of the general contractor's work makes continuity of work for a crew possible. Therefore, constant coordination is mandatory. Nevertheless, the planning schedule must assume that the electrical crews will move from one activity to another, and the Activity Schedule must properly provide for this.

A standard layout should be used for all schedules. Each activity must be numbered, and the number shown in a circle at the end of the schedule. The Activity Schedule represents the ultimate in job management techniques.

Ideally, these charts should be followed exactly, but since the General Contractor controls the progress of the job, it is necessary to be flexible. The alternative is periodic review of the progress of the job and constant revision when necessary.

The Activity Schedule should show references to the incorporation of the special machines, tools, and construction practices. Any special information relative to the the management of the job should also be shown on the schedule. The Activity Schedule is the most important tool available to management, and it must be used to every advantage.

Experience indicates that a minimum of job subdivisions is best for scheduling. However, it would be a serious mistake to fail to take advantage of job subdivisions when they are really necessary.

Each subdivision of a job requires a separate Bill of Material and, in fact, is treated as an individual operation. The principal advantage of subdividing a job is that material does not have to be handled and stored on the job over long periods of time.

Separate buildings or facilities in a multibuilding complex should always be treated as subdivisions. In a multistory building, the basement and first floor with riser takeoff should be separated from upper floors. A physically separated area whose accomplishment has been scheduled late in the job should also be recognized as a subdivision.

Although subdividing a job provides better control for material scheduling, in some cases it can introduce excessive complications for the field personnel, to the point where the advantages disappear.

The chart should be prepared on a piece of transparent paper, 17x22 in., and photoreduced to a size of 8½x11 in. for optimum filing and reference.

ACTIVITY SCHEDULE

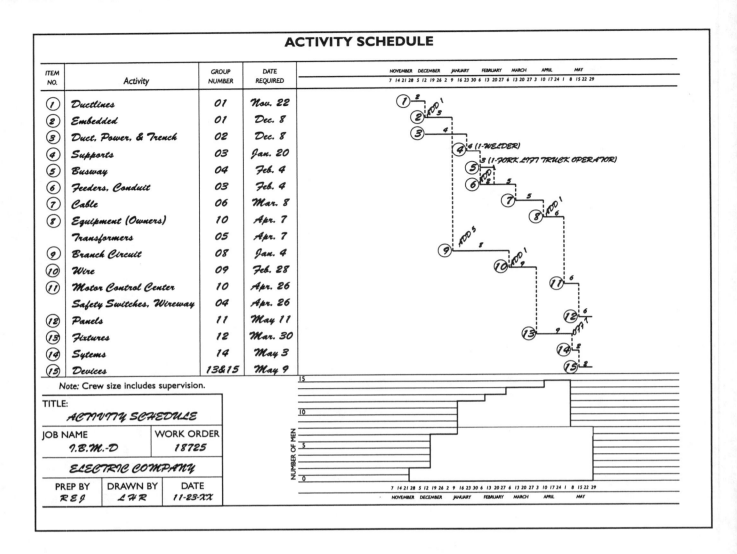

ITEM NO.	Activity	GROUP NUMBER	DATE REQUIRED
1	Ductlines	01	Nov. 22
2	Embedded	01	Dec. 8
3	Duct, Power, & Trench	02	Dec. 8
4	Supports	03	Jan. 20
5	Busway	04	Feb. 4
6	Feeders, Conduit	03	Feb. 4
7	Cable	06	Mar. 8
8	Equipment (Owners)	10	Apr. 7
	Transformers	05	Apr. 7
9	Branch Circuit	08	Jan. 4
10	Wire	09	Feb. 28
11	Motor Control Center	10	Apr. 26
	Safety Switches, Wireway	04	Apr. 26
12	Panels	11	May 11
13	Fixtures	12	Mar. 30
14	Sytems	14	May 3
15	Devices	13&15	May 9

Note: Crew size includes supervision.

TITLE:
ACTIVITY SCHEDULE

JOB NAME	WORK ORDER
I.B.M.-D	18725

ELECTRIC COMPANY

PREP BY	DRAWN BY	DATE
REJ	LHR	11-23-XX

ACTIVITY SCHEDULE

ITEM NO.	Activity	GROUP NUMBER	DATE REQUIRED

Calendar header (months and weeks):

NOVEMBER				DECEMBER			JANUARY				FEBRUARY			MARCH			APRIL			MAY		
7	14	21	28	5	12	19	26	2	9	16	23	30	6	13	20	27	6	13	20	27	3	10	17	24	1	8	15	22	29

NUMBER OF MEN

15

10

5

0

Calendar footer (months and weeks):

NOVEMBER				DECEMBER			JANUARY				FEBRUARY			MARCH			APRIL			MAY
7	14	21	28	5	12	19	26	2	9	16	23	30	6	13	20	27	6	13	20	27	3	10	17	24	1	8	15	22	29

Note: Crew size includes supervision.

TITLE:

JOB NAME		WORK ORDER	
PREP BY	DRAWN BY	DATE	

NAME	**BILL OF MATERIAL**
NUMBER	**JP-5**
ORIGINATOR	Job Processor
COPIES	5
DISTRIBUTION	Purchasing Department, packing ticket, Contract Manager, hold copy, field copy
SIZE	8½ × 11 in.
PURPOSE	All the items prepared by the Job Processor are important for the welfare of the job, but the single most important document to the welfare of the overall operation of the electrical contractor's business is the Bill of Material. Its preparation cannot be undertaken until the following processes have been completed:

✓ Complete checkout of the electrical blueprints
✓ Preparation of blueprints to show feeder routing
✓ Determination of types of construction to be used by General Contractor, and select types of assemblies required
✓ Layout of branch circuit work
✓ Preparation of fixture schedule
✓ Preparation of panelboard schedule
✓ Preparation of Activity Schedule for material requirement dates

The Bill of Material is divided according to Work Schedules, which in their simplest form are:

✓ Embedded work.   ✓ Conductor.
✓ Feeder conduit work.   ✓ Lighting fixtures.
✓ Equipment.   ✓ Finishing.
✓ Branch circuit work.

The Work Schedules must not only be practical subdivisions for easy reporting, but must also be keyed to the subdivisions of electrical construction work because the estimator must prepare cost estimates for every Work Schedule.

Every effort must be made to prepare a Bill of Material that is perfect in every detail, but this is very difficult for many reasons. The most important characteristics that a usable Bill of Material must have are the following:

✓ It indicates the right quantities.
✓ It gives the right catalog numbers and a good description of the items, including the color, finish, or size.
✓ It shows the dates required.
✓ The requirement dates are those shown on the Activity Schedule.

Each Work Schedule material list must be listed separately on the Bill of Material.

BILL OF MATERIAL

Floor Number: _____ Group: _____ Lead Time: _____ Sheet Number: _____

Job Name: _____ Ship to: _____ Date: ____ / ____ / ____

Contract Number: _____ Engineer: _____ Contract Manager: _____ Work Order Number: ____

Quantity Shipped	Quantity Ordered	C L	Item Code	Description	Where Used	Date Required	P.O. Number

NAME	**JOB EXPENSE REQUISITION**

NAME **JOB EXPENSE REQUISITION**

NUMBER **JP-6**

ORIGINATOR **Job Processor**

COPIES **5**

DISTRIBUTION **Purchasing Department, packing ticket, Contract Manager, hold copy, field copy**

SIZE **8½ × 11 in.**

PURPOSE The services required for a job outside the company's operations are ordered on a Job Expense Requisition.

This form is used in place of a Bill of Material whenever the expense is a direct job expense and is not a part of the material cost. The expense is handled in this way for accounting reasons. Typical job expense items are:

✓ Heavy hauling.

✓ Crane service.

✓ Trenching.

✓ High-potential testing.

✓ Concrete sawing.

✓ Core drilling.

✓ Telephone services.

✓ Carpentry work.

✓ Painting.

✓ Sound system installations.

If the work is extensive and involved, a subcontract agreement should be prepared. When arrangements for the work are made by telephone and a Purchase Order number given to the vendor, the Job Expense Requisition should be prepared also, and a notation made on the form "confirming." The Job Expense Requisition is distributed and filed in the same manner as the Bill of Material.

JOB EXPENSE REQUISITION

Floor Number:_____ Group:_____ Lead Time:_____ Sheet Number:_____

Job Name:_____ Ship to:_____ Date:____ / ____ / ____

Contract Number:_____ Engineer:_____ Contract Manager:_____ Work Order Number:_____

Description	Where Used	Date Required	P.O. Number

Amount of Contract:_____

NAME	**FIXTURE SCHEDULE**
NUMBER	**JP-7**
ORIGINATOR	Job Processor
COPIES	4
DISTRIBUTION	Contract Manager, Construction Manager, field, Purchasing Department
SIZE	8½ × 11 in.

PURPOSE   The journeyman on the job must have complete information on everything pertaining to lighting fixtures on the job. This is accomplished by preparing a Fixture Schedule.

The following information is shown on this form:

- ✓ Quantity of fixtures on job by type
- ✓ Information on lamps in designated fixture, including number of lamps per fixture; type, size, or shape of lamps; color of lamps; wattage of individual lamps
- ✓ Description of each type of fixture
- ✓ Finish of fixture
- ✓ Mounting or type of hanger
- ✓ Manufacturer's name
- ✓ Catalog number
- ✓ Location (names of several locations, if necessary)

A horizontal line is drawn separating the fixture types after the data on each type of fixture are complete.

When the schedule is complete, it is incorporated into one of the electrical drawings for use in the field; this is done by typing the Fixture Schedule on tracing paper and cutting it into the regular tracing.

FIXTURE SCHEDULE

Job Name: _____   Work Order Number: _____

Quantity	Type	Lamps				Description	Finish	Mounting	Manufacturer	Catalog Number	Location
		No.	Type	Color	Watts						

NAME	**PANELBOARD SCHEDULE**
NUMBER	**JP-8**
ORIGINATOR	**Job Processor**
COPIES	**4**
DISTRIBUTION	**Contract Manager, Construction Manager, field, Purchasing Department**
SIZE	**8½ × 11 in.**

PURPOSE   The Panelboard Schedule lists the circuits and the loads they serve in any one panelboard. The completed Panelboard Schedule gives all the electrical information relative to the character of the panelboard. Good design would suggest that the design KVA equal or exceed the connected KVA. Under these conditions, the demand and diversity factor represent the spare capacity of the panelboard. Most panel manufacturers number the breakers with odd numbers on the left side and even numbers on the right side.

    The Panelboard Schedule is arranged in such a way as to avoid having two circuits on the same phase, in the same conduit with a common neutral; this is because the circuit numbers in any conduit must be either all odd or all even—not a combination of both. For each circuit, the following should be shown:

✓ Circuit number

✓ Number of poles and amperage of circuit (In the case of 2- or 3-pole breakers, two or three circuit numbers are used, and one or two lines are blank.)

✓ Circuit description (The load is described as the number of fixtures, receptacles, motors, etc., on that circuit.)

✓ Description of the location of the circuit area

✓ Approximate connected load, using 200 VA per duplex receptacle

    The Panelboard Schedule is used in circuiting the job. The circuit information is listed as the circuit numbers are assigned. After the circuits have all been listed and the total connected load has been determined, the feeder may be sized.

    When the Panelboard Schedules have been completed, they should be incorporated into the electrical drawings for use in the field. This is done by typing the schedules on tracing paper and cutting them into the regular tracing.

    If the panelboard is connected exactly as shown on the schedule, the panelboard directory can be typed directly from the schedule. Since this procedure will save considerable time in the field, every effort should be made to connect the panelboard according to the schedule shown on the plans as prepared by the Processor. The Processor's work is extremely important in effecting a smooth running operation in the field and office of the electrical contractor, and nothing should interfere with its early completion after the contract has been signed.

PANELBOARD SCHEDULE

Connected KVA: _____

Job Name: _____

Design KVA: _____

Panel: _____ Type: _____ Mounting: _____ Voltage and Phase: _____

Location: _____ Feed: _____ In: _____ Main Lug Location: _____

Circuit Number	Breaker AMP/Pole	Circuit Description	Location	Watts		Circuit Number	Breaker AMP/Pole	Circuit Descripion	Location	Watts
1						2				
3						4				
5						6				
7						8				
9						10				
11						12				
13						14				
15						16				
17						18				
19						20				
21						22				
23						24				
25						26				
27						28				
29						30				
31						32				
33						34				
35						36				
37						38				
39						40				
41						42				

NAME	**WORK ORDER**
NUMBER	**JP-9**
ORIGINATOR	**Construction Manager or Dispatcher**
COPIES	**3**
DISTRIBUTION	**Job Processor, Purchasing Department, Accounting, or Dispatcher**
SIZE	**8½ × 11 in.**
PURPOSE	This form provides the information pertinent to the job to be installed as follows:

1. Complete information relative to the name and address of the customer, indicating the name of the person who ordered the work and showing P.O. number
2. Complete description of work to be done
3. Complete description of work done
4. Tabulation of material items, with provisions for pricing
5. Tabulation of labor hours worked, and provision for reporting the equipment used
6. Space for summarizing the customer cost of material, equipment, and labor, showing the total cost to the customer

The Work Order is the basic document for every job. Its purpose is to authorize work in the field and to inform other departments of the contractual conditions of the job.

To account for each Work Order number and to ensure against duplications and omissions, a log of issued Work Order numbers is maintained in numerical order. The construction secretary issues the Work Orders and records the numbers in the log with their associated job name and Contract Manager. The log is to be kept in the construction secretary's file drawer, along with the Work Order forms. The construction manager or the processor establishes the Work Order number and the job name for each job on the day the job file is received. The construction manager obtains the Work Order from the construction secretary.

A Request for Work Order is handwritten by the processor, and that information is typed on the Work Order by the construction secretary.

The Work Order is typed in triplicate. The billing copy is transmitted to data processing for entry in the Work Progress Report. The field copy is transmitted to the field. Upon completion of the work, the Foreman signs that copy, dates it, and sends it to the Contract Manager. When this copy is sent to Accounting, it signals that the job is completed and that final billing should be issued. The engineering copy is filed in the contract data folder.

The Work Order establishes the job throughout the organization, and therefore it is vital that it be written and distributed as soon as possible after the job is received. It is initiated as a processing function.

WORK ORDER

Date	Type	Job Number	Work Order Number

Work Location	Billing Address

Customer P.O.	Ordered by	Telephone Number	

Description of Work Requested

_____
_____
_____
_____
_____

Description of Work Done:

_____
_____
_____
_____

Quantity	Material Description	Price	Amount	Quantity	Material Description	Price	Amount

Date	Employee Number	S/T Hours	T/H Hours	D/T Hours	Equipment Number	Material	S
						Equipment	$
						Hrs. @	$
						Hrs. @	$

AUTHORIZATION: I hereby state that I am the (owner) (authorized representative of the owner) and authorize Electric Company, Inc. to perform the above requested work. I understand and agree to pay for said work immediately upon its completion, unless prior arrangements have been made. Should I fail to pay as agreed, I will be obligated to prorate charges in the amount of 1½% per month on the unpaid balance. I further agree that in the event legal proceedings are instituted to collect the unpaid balance, I will pay all costs incurred in connection with said proceedings, including a reasonable attorney's fee.

Please pay this amount:	$

Authorized Signature: _____ Title: _____

Billing Instructions, Add to P.O./Contract Number: _____

CHAPTER NINE
CLAIMS

FORM NO.	FORM NAME	PAGE
C-1	Table of Overtime Premium Pay Inefficiency	256
C-2	Inefficient Labor Due to Acceleration	258
C-3	Summary of Costs for Acceleration	260
C-4	Engineering Clarification Request	262
C-5	Work Stoppage Report	264
C-6	Summary of Costs for Engineering	266
C-7	Final Summary of Direct Job Costs	268
C-8	Allocation of Home Office Indirect Expense	270
C-9	Extended and Excessive Overhead	272
C-10	Reimbursement of Cost of Borrowed Capital	274
C-11	Man-Hours Lost Due to Overloading	276
C-12	Overload Condition Evaluation	278
C-13	Price Proposal for Overloading	280
C-14	Time Extension Proposal	282

As a construction job progresses, undesirable working conditions may arise, which are created by others and are not controllable by the contractor. These conditions usually result in additional costs because they create a loss of productivity of the workers. It is vitally important for contractors and their personnel to recognize the conditions that cause the losses and, further, to be able to evaluate them.

These unusual conditions result in claims against the party (General Contractor, architect, engineer, or owner) who caused them, and consist of the following types:

1. Acceleration
2. Engineering
3. Financing
4. Joint occupancy
5. Extended overhead and job expense
6. Specification interpretation
7. Impact

Individually or collectively, these claims can greatly increase costs to the contractor. If proper recognition is to be given to each of these factors as the job progresses, the contractor and everyone in the organization must be familiar with them and properly document them.

NAME
NUMBER
ORIGINATOR
COPIES
DISTRIBUTION
SIZE

PURPOSE

TABLE OF OVERTIME PREMIUM PAY INEFFICIENCY

NUMBER C-1

ORIGINATOR Construction Manager

COPIES 2

DISTRIBUTION General Manager, Accounting

SIZE 8½ × 11 in.

PURPOSE With this form, the contractor tabulates the overtime work performed on the job according to the total hours worked per week. To this tabulation of man-hours is applied the work efficiency percentage to arrive at the total hours lost for acceleration. Overtime was incurred primarily as a result of acceleration directives issued to the Prime Contractor, who in turn passed on such directives to the subcontractor. As shown in the example below, approximately 95% of all overtime occurred after, and as a result of, the acceleration directives.

TABLE OF OVERTIME PREMIUM PAY INEFFICIENCY

Job Name: _____   Date: ___ / ___ / ___

Month	Hours Worked	Overtime Hours	Six 8-Hour	Seven 8-Hour	Six 10-Hour	Seven 10-Hour
June		22	3			
July		25	3			
August		66	9			
September		24	3			
October		23	3			
November		703	86			
December		212	27			
January		–	–			
February		21	3			
March		246	31			
April		2,505	313			
	102,402	3,847	481			
May		4,600	100	–	65	88
June		5,600	59	2	152	109
July		758	1	8	–	6
August		4,062	–	–	195	–
September		6,920	11	9	307	–
October		7,158	14	3	209	163
November		811	35	–	–	–
December		367	16	–	–	–
	92,165	30,276	236	22	928	366
Total Hours:	194,567	34,123	717	22	928	366

The Effect of Overtime

A study of the effect of overtime work on labor productivity reveals three things:

1. A nominal amount of overtime performed occasionally causes only a small decrease, if any, on productivity.
2. More than a nominal amount of overtime, regularly required, definitely decreases productivity.
3. When there is an urgent need of completion of a job or sections of a job, production during the overtime period may increase, but nearly always decreases.

The following is a summary of the results of this study:

Table of Overtime Efficiency Rates

Hours Worked	How Worked	Overtime Applicable to	Number of Overtime Hours	Efficiency Rates (%)
Five 10-hour days	On same job	Hours over 8	2	87.5
Five 12-hour days	On same job	Hours over 8	4	75.0
Five 10-hour days	Workers moved to other job after 8 hours	Hours over 8	2	75.0
Five 12-hour days	Workers moved to other job after 8 hours	Hours over 8	4	68.8
Six 8-hour days	On same job	Saturday only	8	87.5
Six 10-hour days	On same job	Total hours	60	84.5
Seven 8-hour days	On same job	Saturday and Sunday	16	77.0
Seven 10-hour days	On same job	Total hours	70	77.8

SOURCE: Efficiency rates from NECA.

TABLE OF OVERTIME PREMIUM PAY INEFFICIENCY

Job Name: _____   Date: ____ / ____ / ____

Month	Hours Worked	Overtime Hours	Six 8-Hour	Seven 8-Hour	Six 10-Hour	Seven 10-Hour
_____						
_____						
_____						
_____						
_____						
_____						
_____						
_____						
_____						
_____						
_____						
_____						
_____						
_____						
_____						
_____						
_____						
_____						
_____						
_____						
_____						
_____						
_____						
_____						
_____						
_____						
_____						
_____						
_____						
_____						
Total Hours:						

NAME	**INEFFICIENT LABOR DUE TO ACCELERATION**
NUMBER	**C-2**
ORIGINATOR	**Construction Manager**
COPIES	**2**
DISTRIBUTION	**General Manager, Construction Manager**
SIZE	**8½ × 11 in.**

PURPOSE When an electrical contractor successfully bids on a job whose construction schedule is based on a normal completion time, it is appropriate to assume that the work will be done under normal conditions. If the owner then decides in the course of construction that the completion date must be advanced, the electrical contractor is faced with two alternatives:

1. Increase the number of workers on the job.
2. Work the existing crew overtime to keep up with the increased work schedule.

In either case, the job is subjected to *acceleration,* and the contractor is exposed to increased costs for which reimbursement is due.

To justify a claim for the increased cost, it is necessary for the contractor to illustrate the effect the changed condition had on the manpower loading by comparing the manpower chart prepared at the beginning of the job with the manpower chart of the actual work done under the accelerated conditions. Acceleration nearly always results in overtime, and overtime work results in reduced productivity due to fatigue and change of attitude. The extent of loss of efficiency varies according to the length of the work day and the number of days worked during the week.

INEFFICIENT LABOR DUE TO ACCELERATION

Job Name: _____   Date:___ / ___ / ___

Number of Man-Weeks Worked	Weekly Hours Worked	Weekly Hours of Lost Efficiency	Total Hours of Lost Efficiency	Unit Percent Efficiency Loss	Hours Lost	Cost of Labor Loss @$15.00/M.H.
400	Five 10-hour days	10	4,000	12.5	500	$ 7,500
100	Six 8-hour days	8	800	12.5	100	$ 1,500
100	Six 10-hour days	60	6,000	15.5	930	$13,950
200	Seven 8-hour days	16	3,200	23.0	730	$11,040
100	Seven 10-hour days	70	7,000	22.2	1,554	$23,310
Totals			21,000		3,820	$57,300†

Basis of typical schedule:
· *Man-week assumed.*
· *Efficiency rates from NECA reports.*
· *Efficiencies of 8-hour days apply to Saturday and Sunday.*
· *Efficiencies of 10-hour days over 5 days apply to total hours.*
· *Efficiency of 10-hour days on 5-day week apply to hours over 8.*
† *Total loss is $57,300 plus fringes.*

INEFFICIENT LABOR DUE TO ACCELERATION

Job Name: _____   Date:_____ / _____ / _____

Number of Man-Weeks Worked	Weekly Hours Worked	Weekly Hours of Lost Efficiency	Total Hours of Lost Efficiency	Unit Percent Efficiency Loss	Hours Lost	Cost of Labor Loss
Totals						

NAME	**SUMMARY OF COSTS FOR ACCELERATION**
NUMBER	**C-3**
ORIGINATOR	**Construction Manager**
COPIES	**2**
DISTRIBUTION	**Construction Manager, General Manager**
SIZE	**8½ × 11 in.**
PURPOSE	On this form, the Construction Manager can summarize excessive costs due to acceleration.

SUMMARY OF COSTS FOR ACCELERATION

Job Name: _N.A.D.C._   Date: _12_ / _15_ / _XX_

Item	Unit	Detail	Detail Total	Detail Total
Total hours worked overtime	21,000			
Efficiency loss hours	3,820			
Cost of labor loss	$15/Hour	15 × 3,820	$57,300	
Cost of supervision	15%/Lab.	.15 × $57,300 hrs.	8,595	
Total labor base cost				$65,895
Cost of fringes	31% Lab.	.31 × 65,895 hrs.	20,427	
Cost of small tools and supplies	7% Lab.	.07 × 65,895 hrs.	4,612	25,039
Total net labor cost				$90,934
Job overhead cost	15% Net	.15 × 90,934	13,640	
Home office cost	4.5% Net	.045 × 90,934	4,092	
Total prime cost			$108,666	
Profit	10% Prime	.10 × 108,666		10,866
Total billing for acceleration:				$119,532

SUMMARY OF COSTS FOR ACCELERATION

Job Name: _____   Date:____ / ____ / ____

Item	Unit	Detail	Detail Total	Summary Total
Total hours worked overtime	_____			
Efficiency loss hours	_____			
Cost of labor loss	____ /Hour	____ x ____		
Cost of supervision	____ %/Lab.	____ x ____ hrs.	_____	
Total labor base cost				_____
Cost of fringes	____ % Lab.	____ x ____ hrs.	_____	
Cost of small tools and supplies	____ % Lab.	____ x ____ hrs.	_____	_____
Total net labor cost				
Job overhead cost	____ % Net	____ x ____	_____	
Home office cost	____ % Net	____ x ____	_____	
Total prime cost				
Profit	____ % Prime	____ x ____		

Total billing for acceleration: _____

NAME	**ENGINEERING CLARIFICATION REQUEST**
NUMBER	**C-4**
ORIGINATOR	Job Foreman
COPIES	2
DISTRIBUTION	Job Foreman, Project Manager
SIZE	8½ × 11 in.
PURPOSE	With this form, the Job Foreman can request a change of job blueprints due to orders from the General Contractor to modify the work.

With this form, the Job Foreman can request a change of job blueprints due to orders from the General Contractor to modify the work.

The cost of engineering work directly connected with each change should be claimed separately. In the case of settled modifications where additional direct engineering is claimed and paid in full, such payment is credited to the claim. However, in addition to the direct engineering effort, labor, and material that can be isolated and attributed to a given change, other factors directly and indirectly affect both total costs and the time required for performance, including:

✓ The adverse effect of changes on the overall labor and engineering effort in terms of disruptions in previously planned work schedules, the necessity to hold up phases of the job pending approval of a change, and the attitude of workers on the job.

✓ The cost of estimating the price of the changed work and the time required to perform it, including time spent searching through previous modifications to make sure that the latest contract requirements are considered.

✓ The cost of travel and sometimes lengthy negotiations covering a multitude of separate items within a given modification.

✓ The cost of extra layout and preparation of working drawings.

✓ The cost of additional material approval submissions.

✓ The cost of extra clerical work for purchasing additional materials, maintaining cost records, amending subcontracts, and billing for the changes.

✓ The cost of additional financing required to do more work within a given time.

For the purposes of a claim, *impact costs* are defined as increased costs that are attributable to the cumulative effect of numerous changes in the original contract, and to the resultant disruptions and inefficiencies introduced into a contractor's construction operations. Impact costs result from factors such as mobilization, demobilization, time spent waiting for problems to be resolved, interruptions to overall planning and scheduling, and inefficiency due to lack of continuity of the work.

Claims for additional engineering hours are based on the concept that, had the contract drawings and specifications been adequate, it would have been possible to greatly reduce the engineering effort at about the time the construction work was getting fully underway, while instead it was necessary to continue a heavy engineering effort right up to the very end of construction.

ENGINEERING CLARIFICATION REQUEST

By: _____ Title:_____ Date:____ / ____ / ____

Engineering Clarification Is Needed as Follows:

Area:_____ Location in Area: _____

_____

_____

Drawing Reference:_____ Work Order Number: _____

Remarks: (Indicate all specific trouble areas and drawing discrepancies.)

_____

_____

_____

_____

_____

_____

_____

_____

_____

_____

_____

_____

_____

_____

_____

_____

_____

_____

_____

_____

_____

_____

_____

_____

_____

NAME	**WORK STOPPAGE REPORT**
NUMBER	**C-5**
ORIGINATOR	**Job Foreman**
COPIES	**2**
DISTRIBUTION	**Job Foreman, Project Manager**
SIZE	**8½ × 5½ in. (Copy form at 100% of original; cut it out along the box rules.)**
PURPOSE	When work is stopped in anticipation of changes or as a result of construction difficulty, this form is used to notify the Project Manager and to identify the extent of the loss of time.

The costs of items to change orders are usually *far in excess* of the prorated costs for an entire job which is performed with no, or only a few, change orders. Since these are costs due to a specific action on a specific job, it is more proper to treat them as direct job expense rather than overhead. Such costs are in addition to a contractor's normal operating or overhead costs related to his total volume of business.

The situation where the customer or customer's representative attempts to evaluate the cost of change orders only in unit proportion to comparable items of costs related to an entire job involves an improper practice.

In the case of electrical construction work, the tendency of some customers or customer's representative to maintain arbitrary limits on allowable general operating costs or overhead at the level of such costs for general contractors is an improper practice. *Such costs are considerably higher*, percentagewise, for electrical and other similar specialty trade *contractors*.

The amount of extra engineering effort and related costs increases in proportion to the number of contract modifications that have been directed and the number of necessary changes that have been discovered by the contractor. Also, the impact of changed work on the original basic contract work is greater during periods of concentrated effort (acceleration) than it is during normal periods. The impact of changes increases in proportion to the rate of changes and is related to the stages of construction during which the changes are made.

Sometimes a substantial additional, unforeseen engineering effort is required because of errors and ambiguities in the contract drawings, the necessity to resolve design deficiencies, changes to the contract work and changes to changes, discrepancies in drawings and specifications for equipment, late delivery of drawings and specification equipment, and nonconformance of, and defects in, equipment. A tremendous volume of correspondence is required to clarify and correct errors, ambiguities, and omissions in contact drawings and specifications.

In the field, two forms should be used to record engineering deficiencies: the Work Stoppage Report (Form C-5) and the Engineering Clarification Request (Form C-4).

WORK STOPPAGE REPORT

Area: _____

Date Work Stopped: _____ Time: _____

Date Work Resumed: _____ Time: _____ Total Man-hours Delay: _____

Description of Work Stopped: _____

_____

_____

Reason Work Stopped: _____

_____

_____

_____

_____

_____

_____

Responsibility: ☐ Engineer ☐ General ☐ C.O.E. _____
(Other)

Signed: _____ Date of Report: _____

NAME SUMMARY OF COSTS FOR ENGINEERING

NUMBER **C-6**

ORIGINATOR **Project Manager**

COPIES **2**

DISTRIBUTION **Construction Manager, Accounting**

SIZE **8½ × 11 in.**

PURPOSE This form enables the Project Manager to report a claim for additional engineering costs required because of field changes in the work, and determine these costs. As used in this claim, *impact costs* are the increased costs attributable to the cumulative effect of numerous modifications coming one on another and on the original contract work, as well as to delays, disruptions, and inefficiencies introduced into the construction operations through no fault of the contractor.

SUMMARY OF COSTS FOR ENGINEERING

Job Name: _D.I.A._   Date: _10 10 94_

Item	Unit	Detail	Detail Total	Summary Total
Hours lost/impact on man-hours	3500			
Work stoppage report	1200			
Hours lost by engineering change	400	5100		
Cost of labor loss	$15/hour	5100 × $15	76500	
Cost of supervision	15%/ labor	0.15 × 5100 hours	11475	87975
Total labor base cost				
Cost of fringes	31% labor	31 × 5100 hours	23750	
Cost of small tools and supplies	7% labor	0.07 × 5100 hours	5355	29105
Total net labor cost				119080
Engineering cost	310 hours	15 × 310 hours	4650	121730
Job overhead cost	15% net	0.15 × 3	18259	
Home office cost	4.5% net	0.045 ×	5477	23736
Total prime cost				145466
Profit	10% prime	0.10 ×		14546
		Total		$160012

Total Billing for Engineering: _____

SUMMARY OF COSTS FOR ENGINEERING

Job Name: _____   Date:____ / ____ / ____

Item	Unit	Detail	Detail Total	Summary Total
Hours lost/impact on man-hours	_____			
Work stoppage report	_____			
Hours lost by engineering change	_____	_____		
Cost of labor loss	/hour	×	_____	
Cost of supervision	15%/labor	0.15 × hours	_____	
Total labor base cost				_____
Cost of fringes	____% labor	× hours		
Cost of small tools and supplies	7% labor	0.07 × hours	_____	_____
Total net labor cost				_____
Engineering cost	hours	× hours		_____
Job overhead cost	15% net	0.15 ×	_____	
Home office cost	4.5% net	0.045 ×	_____	
Total prime cost				_____
Profit	10% prime	0.10 ×		_____

Total Billing for Engineering: _____

NAME	**FINAL SUMMARY OF DIRECT JOB COSTS**
NUMBER	**C-7**
ORIGINATOR	**Project Manager**
COPIES	**2**
DISTRIBUTION	**Construction Manager, Accounting**
SIZE	**8½ × 11 in.**
PURPOSE	This form contains a summary of all of the actual direct cost on the overall job, including subcontracted costs and all other costs directly chargeable to the job.

FINAL SUMMARY OF DIRECT JOB COSTS

Job Name: _____   Date:____ / ____ / ____

Material cost		
Labor cost		
Subcontracts		
Supervision		
Engineering labor		
Clerical labor		
Labor fringe benefits		
Subsistence		
Performance bond		
Equipment costs		
Outside engineering		
Testing		
Outside services		
Office expense		
Travel expense		
Expendable tools		
Field supplies		
Total direct costs		

NAME	**ALLOCATION OF HOME OFFICE INDIRECT EXPENSE**
NUMBER	**C-8**
ORIGINATOR	**Construction Manager**
COPIES	**2**
DISTRIBUTION	**Construction, Manager, Accounting**
SIZE	**8½ × 11 in.**

PURPOSE   With this form, the Construction Manager can determine the amount of the home office overhead chargeable to the job being done remotely from the home office.

Indirect expense includes overhead expense and general and administrative expense, as well as the cost of maintaining the home office facility.

The evaluation is made by determining the ratio of direct costs of the home office during the period the job was underway to the direct costs of the job. The percentage of the home office indirect expenses is then applied to the costs of the job.

This home office expense is an allowable job expense because the home office performs so many job operations, such as payroll, material purchasing and delivery, top supervision of the job, and the like.

ALLOCATION OF HOME OFFICE INDIRECT EXPENSE

Job Name: _____   Date:____ / ____ / ____

Duration of job in months

Total home office direct costs

Total job direct costs (electrical)

Total job subcontracts

Total job costs

Percent home office of job costs

Home office indirect costs/duration

Total home office indirect costs

Percent of home office indirect costs

Home office indirect expense allocable to job

Submitted by: _____   Date:____ / ____ / ____

NAME **EXTENDED AND EXCESSIVE OVERHEAD**

NUMBER **C-9**

ORIGINATOR **Comptroller**

COPIES **3**

DISTRIBUTION **Claims Report, General Manager, Comptroller**

SIZE **8½ × 11 in.**

PURPOSE *Overhead* is comprised of general operating expenses that are necessary irrespective of whether any given jobs are performed. These are all costs other than those that can be directly charged to a particular project such as direct labor, direct material, subcontracts, and other separable direct job expenses.

Since overhead is partially a product of time, "extended overhead" and "acceleration" might appear to be incompatible. Even if overhead were *entirely* a product of time, however, extended overhead would only be incompatible with acceleration if the original contemplated period of performance was in fact reduced.

Where direct job expense and overhead have been allowed in modifications, such allowances have been credited to this claim. The normal direct job expenses and overhead as contemplated in the bid price have been deducted from actual costs. In a sense, this is a total cost approach. But it should be obvious that it is extremely difficult, if not impossible, to proceed on any other basis, as far as direct job expenses and allowable overhead is concerned. Briefly stated, this suggested basis for arriving at an equitable settlement is the difference between the reasonable costs of the work as bid and the reasonable costs of actual contract performance.

The cases allowing equitable adjustments on a total cost basis require proof that such costs are incurred directly on, or are properly applicable to, the contract in question on an overall basis. The fact that such costs cannot be isolated to a precise segment of the original or changed contract work does not provide a basis for rejection. Once the determination has been made that costs incurred are allowable costs and have been incurred on a given contract, the courts and boards seem willing to allow a total cost settlement, as long as the costs to be compared (i.e., anticipated costs and actual costs) are reasonable.

EXTENDED AND EXCESSIVE OVERHEAD

Job Name:_____ Date:____ / ____ / ____

Account	Item	Detail	Total
Total indirect expense			
Original estimated overhead			
Less: Overhead in approved change orders			
Total extended overhead			
Total project direct job expense			
Less: Subcontracts			
Actual total job expense			
Estimated job expense			
Job expense overrun			
Less: Direct job expense approved in change order			
Net extended job expense			
Net overrun overhead and direct job expense			
Profit at 10%			
Total price			

NAME	**REIMBURSEMENT OF COST OF BORROWED CAPITAL**
NUMBER	**C-10**
ORIGINATOR	**Comptroller**
COPIES	**3**
DISTRIBUTION	**General Manager, Claims Report, Comptroller**
SIZE	**8½ × 11 in.**
PURPOSE	With this form, the Comptroller determines the cost of borrowed capital required to finance the additional work on a job, over and above the contract amount.

When planning the volume of business required to satisfy the budget, and to conform with the available operating capital of the business, it is necessary to include some small and some large jobs in order to arrive at a suitable "mix." However, if in the course of the year's work some of the jobs involve a disproportionately large number of Change Orders of substantial magnitude, the contractor may have to supplement working capital with bank loans to meet the financial requirements of the work. It is possible for a contract to double in size, with the customer demanding that it be completed within the same time frame specified in the original contract. Needless to say, this will stretch the financial capabilities of most firms when the job is a particularly large one.

Clearly, it takes considerably more working capital to finance a job costing several million dollars than one costing half as much in the same period. Presumably, any financing cost required to perform the original contract will be included in the original bid, and the question at issue will concern only the additional financing charges. While the amount of additional working capital needed may have been increased because of the failure of government to provide proper payments as the work progressed, this cannot be considered to negate the basic premise: that as significant additional work is added within a given performance period there must be a corresponding increase in working capital. As additional working capital is required, additional interest must be paid.

It stands to reason that interest expense for working capital borrowed for the performance of ordinary construction work is not reimbursable when there are no limitations on the time in which the work must be performed. However, when there is no extension of time allowed to perform an amount of work in excess of that specified in the original contract, then interest on additional borrowed working capital is a compensable expense.

REIMBURSEMENT OF COST OF BORROWED CAPITAL

Job Name: _____   Date:____ / ____ / ____

Total final labor cost	$	
Total final material cost		
Total final direct job expense		
Total prime cost		
Profit 4.8%		
Total job price		$
Less: Original contract price		
Amount of contract extension		
Working capital required of ¹⁄₁₀ amount		
Current bank interest rate _____%		

Submitted by: _____

NAME	**MAN-HOURS LOST DUE TO OVERLOADING**
NUMBER	**C-11**
ORIGINATOR	**Construction Manager**
COPIES	**3**
DISTRIBUTION	**General Manager, Claims Report, Construction Manager**
SIZE	**8½ × 11 in.**

PURPOSE This form, in conjunction with the Construction Progress Chart, provides a means of determining the man-hours lost resulting from joint occupancy. The phrase *joint occupancy* is used to indicate that the work was performed under limited-space conditions with construction forces other than your own, creating a working condition that differed from those represented by the bid documents. Such restricted-workspace conditions have negative consequences:

1. The prolonged presence of an excessive number of workers who occupy the same limited space creates a condition for reduced productivity.
2. Normal working conditions are based on having unrestricted floor areas with minimal obstructions to move or work around, including people, unless the specifications indicate otherwise.

When preparing an estimate for bidding on a job, working conditions must be evaluated very carefully because they have considerable bearing on the efficiency that can be expected on the job. *Efficiency* is defined as the capacity to perform a given task in a specified time, at a reasonable cost, and with a minimum of waste. The factor that determines the amount of the joint occupancy claim is the loss of efficiency of the work force due to their inability to work effectively in a confined space with the limitation forced on them by having conflicts beyond those represented by the specifications. Therefore, it is essential to determine the degree of loss of efficiency of the work force, and in this case it is entirely a question of the rate at which construction is completed compared with the rate that was considered ideal under the conditions represented in the specifications.

To determine the loss of efficiency, it is first necessary to prepare a Construction Progress Chart, as shown, where the solid line represents the engineers' progress schedule, the dashed line shows the actual progress, and the dotted line shows the original progress schedule provided by the general contractor resulting from joint occupancy for the entire project.

CONSTRUCTION PROGRESS CHART

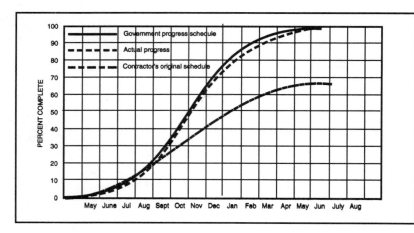

OVERLOAD CONDITION EVALUATION

Job Name: _____   Date:____ / ____ / ____

Month	Monthly Average Number of Workers	Optimum Monthly Worker Loading	Number of Workers Over Optimum	Percent Overload

Average Percent Overload: _____

NAME **OVERLOAD CONDITION EVALUATION**

NUMBER **C-12**

ORIGINATOR **Construction Manager**

COPIES **2**

DISTRIBUTION **Construction Manager, Accounting**

SIZE **8½ × 11 in.**

PURPOSE This form records the monthly manpower volume of workers for the purpose of determining overload conditions on the job, which may be detrimental to achieving maximum productivity.

Optimum manloading is determined by the estimate of total man-hours of work to be completed by an established completion date, divided by the number of man-days of normal man-hours per day.

Overloading is a comparison of actual man-hours worked with the number of optimum workers normally working on each similar day.

From this determination you develop the percent overloading.

As an example, with a project evaluation of 391,000 man-hours to be worked in 13 months, the following optimum man loading was determined:

	Men		Man-Hours
June	50	=	8,800
July	125	=	22,000
Aug.	175	=	30,800
Sept.	225	=	39,600
Oct.	225	=	39,600
Nov.	225	=	39,600
Dec.	225	=	39,600
Jan.	225	=	39,600
Feb.	225	=	39,600
Mar.	225	=	39,600
Apr.	150	=	26,400
May	100	=	17,600
June	50	=	8,800
Total man-hours estimated			391,600

From this projected optimum, a comparison of overloading with reduced productivity efficiency can be made:

MAN-HOURS LOST DUE TO OVERLOADING

Job Name: _____   Date:____ / ____ / ____

Month	Engineers Monthly Progress Schedule	Actual Monthly Progress Schedule	Percent Loss	Efficiency Loss	Man-hours Worked	Monthly Hours Lost

Total hours lost: _____

NAME	**PRICE PROPOSAL FOR OVERLOADING**
NUMBER	**C-13**
ORIGINATOR	**Comptroller**
COPIES	**3**
DISTRIBUTION	**Comptroller, Claims Report, General Manager**
SIZE	**8½ × 11 in.**

PURPOSE On this form, record a total cost for overloading manpower on the job. This condition is brought on by the awarding authority allowing other contractors to work on the job without previously notifying the contractors already on the job.

Too many workers occupying a limited space (as is the case on almost any construction job, but more so on some than others) inescapably results in lower productivity for all the workers.

Two big factors must be figured separately; overtime efficiency loss and overload efficiency loss. The factor of premium pay having been worked is also involved.

PRICE PROPOSAL FOR OVERLOADING

Job Name: _____   Date:____ / ____ / ____

Item of Cost			
Premium pay for overtime worked	_____	_____	
Man-hour overtime efficiency loss	_____	_____	
Man-hour overload efficiency loss	_____	_____	
Total added labor cost		_____	
Less: Acceleration previously charged		_____	
Net labor cost			_____
Supervision—15% labor		_____	
Small tools—7% labor		_____	_____
Prime cost			_____
Job overhead—15%		_____	
Home office overhead—0.045%		_____	_____
Net cost			_____
Profit—10%			_____
Total price			_____

NAME	**TIME EXTENSION PROPOSAL**
NUMBER	**C-14**
ORIGINATOR	**Construction Manager**
COPIES	**3**
DISTRIBUTION	**Construction Manager, Accounting, General Manager**
SIZE	**8½ × 11 in.**

PURPOSE Using this document, the Construction Manager records and determines the time extension for a contract on which there are liquidated damages for the inability to complete the work within a specified period of time.

Liquidated damages may be erased by innumerable Change Orders, causing the addition to the contract of many man-hours of work. All the man-hours are additive, which means that it is not necessary to relate the extension of time to each Change Order.

Working overtime, overtime efficiency loss, and manpower overloading efficiency loss not only influence productivity negatively, but also increase the overall working time.

TIME EXTENSION PROPOSAL

Job Name: _____   Date:_____ / _____ / _____

Month	Working Days per Month	Efficiency Loss Percentage	Days Lost

Total Days Lost: _____

CHAPTER TEN
PURCHASING

FORM NO.	FORM NAME	PAGE
P-1	Purchase Order	288
P-2	Material Schedule Report	290
P-3	Bill of Material for Cable Field Measure	292
P-4	Material Requisition	294
P-5	Request for Proof of Shipment	296
P-6	Information Request Card	298
P-7	Job Expense Form	300

An important factor in achieving productivity in the field is having material, equipment, and tools available when required to make the installation; this is the principal job function of purchasing and warehousing.

Purchasing must look to estimating and job management to list and handle all the items required for the job. Successful material management is therefore a team effort, in which the needs are established by the Estimator, the time required is determined by the Job Processor, Purchasing provides for obtaining the items as needed, and the warehouse makes the delivery in proper containers according to schedule.

At the time of estimating, all the quotations for prices, other than for standard items, are obtained and issued by the Purchasing Department. The Estimator prepares a carbon copy of the pricing sheet and transmits it to Purchasing for quotations. When prices are received, they are noted in the proper column and returned to the Estimator. The Estimator, in turn, notes them on the original pricing sheets and collects the quotation sheets in a separate manila folder.

When the bid is successful, Purchasing receives the quotation sheets back, together with the Bill of Material, which is transmitted for Purchasing according to schedule. In addition to these two documents, revised copies of the Manpower Chart and the Activity Schedule are transmitted as guides for scheduling.

Purchasing receives copies of all pages of the Bill of Material from the Contract Manager, and they are numbered in sequence. This means that the Contract Manager must exercise great care and diligence to see that packaging and handling facilities from the warehouse are available on the job and, furthermore, that the prescribed procedures are followed.

An important part of having material available is expediting it so that requirement dates can be met. This requires meticulous and continuous pursuit. In fact, all the purchasing functions involve precision handling of paperwork, and they must be executed correctly for the achievement of proper scheduling so that material and tools will be available when they are required, which increases productivity.

If the Bills of Material, prepared as a processing function, were 100% accurate as to type of material, quantity needed, and date required, the Contract Manager would have little or no responsibility for material as the job progresses.

While it is the goal to write perfect Bills of Material, it must be recognized that it is impossible to anticipate every material requirement from a set of drawings months before the actual start of a job. Therefore, Contract Managers have a very definite responsibility for material control as the job progresses. As a matter of fact, material control may consume a major portion of their time and be their greatest responsibility, depending on how well the Bills of Material were written in the processing stage.

The Bills of Material can be prepared after (1) the design has been checked and changed (if required), (2) the Proposed Material List has been approved (if required), (3) the general contractor's construction schedule has been received, and (4) the Activity Schedule has been prepared (and approved if required).

The Bill of Material is, without a doubt, the most important document involved in the management of a job. It is the only means of getting material to the job, and it identifies all the material for the entire job by manufacturers' catalog numbers, description, and quantity.

The Bill of Material reflects the planing and scheduling for the entire job, showing the date on which each item is required. It also indicates where items are to be used.

BILL OF MATERIAL FOR CABLE FIELD MEASURE

Floor Number: _____ Group: _Cable field measure_ Lead Time: _____ Sheet Number: _6a_

Job Name: _Empire State Building_ Ship to: _Job-site_ Date: __/__/__

Contr. Number: _____ Engineer: _____ Contract Manager: _V.T._ Work Order Number: _18971_

Quantity Shipped	Quantity Ordered	C L	Item Code	Description		Where Used	Date Required	P.O. Number
	3 pcs.		330423	Wire, Al. 3/0, THW Black	98 ft	LPA	10-31	
				(3-reels)				
	3 pcs.		330423	Wire, Al. 3/0, THW Black	90 ft	PP		39056
	1 pc.		330423	Wire, Al. 3/0 THW Black	57 ft	PP		
				(4-reels)				
	3 pcs.		330425	Wire, Al. 250 MCM, THW Black	132 ft	Service		
	1 pc.		330425	Wire, Al. 250 MCM, THW Black	128 ft	Service		
				(4-reels)				
	3 pcs.		331402	Wire, Cu. #2, THW, Black	73 ft	Elevator		
	1 pc.		331402	Wire, Cu. #2, THW White	78 ft	Elevator		
				(4-coils)				

The effectiveness of the Bill of Material is directly proportional to the accuracy of the original data entered on it and the competence with which instructions contained on it are carried out. The Contract Manager can save himself and many others much time if every effort is made to write perfect Bills of Material. This is difficult to do, and perfection may never be possible 100 percent of the time; however, much improvement can and must be made in the preparation of these documents.

There must always be a description of the item, and it must always begin with the name of the item, followed by the adjectives that describe it, as to where they will be used. This information is available at the time an item is listed, and noting it on the Bill of Material in the Where Used column will save time in the field.

The date required is one of the three most important pieces of information appearing on the Bill of Material; unfortunately, it is the most difficult to get right. The date an item is required is obtained directly from the Activity Schedule, allowing a few days ahead of the actual need for the item.

For Bills of Material for Change Orders or Field Orders, the date an item is required is determined by the anticipated need for it and by the job requirements. This date must be as realistic as possible so that Purchasing will have as much time as possible to deliver the material. Making unnecessary demands on Purchasing as a result of using unrealistic requirement dates is an indication of poor contract management, and must be avoided.

All Bills of Material must be expedited as a function of Purchasing. This assures either the delivery of the material to the job on the date it is required or a notification that the date cannot be met and the establishment of one that can. This communication is effected by expediting information.

NAME	**PURCHASE ORDER**
NUMBER	**P-1**
ORIGINATOR	**Purchasing Agent**
COPIES	**5**
DISTRIBUTION	**Purchasing, Warehouse, Expediting, Accounting**
SIZE	**8½ × 11 in.**

PURPOSE   The Purchase Order enables the Purchasing Agent to record purchasing information, and extend authority to the vendor to furnish materials as requested.

The Purchase Orders for all items of factory-ordered material for each job should be placed as soon as possible for delivery in accordance with scheduled dates issued by the Job Processor. All work schedule materials are processed by the Expediting Clerk after an appropriate lead time has been established for each group. From this information, the Material Schedule Report (Form P-2) is prepared. All groups of material scheduled for initial release to the job are disbursed by the warehouse without a scheduled date, while all the others are set up on a requirement date.

While the Contract Manager is responsible for scheduling and verifying requirement dates for the material delivery, it is the Purchasing Department's responsibility to place the material on the job at the designated time.

Fundamental to all procedures in the Purchasing Department is the proper preparation of the Purchase Order, showing all the information essential for the smooth performance of material handling.

✓ *Show in body of form:* quantity, description, shipping date, cost, requirement date, how shipped, FOB point, Bill of Material number, terms, job name, Work Order number, account number, delivery point

✓ Signature of buyer and shop drawing schedule

The following are important considerations in regard to Purchase Orders:

✓ The order form must be completely filled out.

✓ Separate orders should be written for each requirement date.

✓ Back orders must be avoided, even if the order must be placed with multiple vendors.

✓ Separate Purchase Orders should be written for the items on each work schedule.

✓ At least two quotes must be obtained, and the price paid must conform to that used in the estimate.

✓ Purchasing must meet requirement dates.

PURCHASE ORDER

Name: _____

Address: _____

Telephone: _____

To: _____

Address: _____

Number: _____

Date: _____

Quantity	Standard Number	Description	Price	Per	T/D	Shipping Date

Notification to Vendor:

The attached Purchase Order requires the following paperwork procedures:

1. Manufacturer's Acknowledgment
   It is mandatory for us to have immediate acknowledgment of this purchase showing:
   a. Scheduled shipping date
   b. Vendor's Purchase Order number
   c. Company's Purchase Order number
   d. Job name
   e. Shipping address
2. Shipping Schedule
   We have established an intelligent requirement date for delivery to the job. We must have your cooperation in meeting this date. Therefore,

your shipping date must allow sufficient time to ensure delivery as specified.
3. Shop Drawings (when required)
   These must be available to us as soon as possible, but not later than the date specified, and we must return them promptly. Unnecessary delay of shop drawings might affect shipment.
4. Material Identification
   It is necessary for the manufacturer to show Purchase Order identification such as: Type A fixture, or LPB for panel.
   Your cooperation with these requirements is necessary to ensure proper service.

PLEASE RENDER ALL INVOICES IN TRIPLICATE ON DAY OF SHIPMENT.

Ship via		F.O.B. Point	Req. by	B/M Number	Terms
Job Designation			Work Order Number	Account Number	
Deliver To:					
		By: _____			

A COPY OF SHIPPING OR DELIVERY TICKETS SHOULD ACCOMPANY INVOICE.

NAME	**MATERIAL SCHEDULE REPORT**
NUMBER	**P-2**
ORIGINATOR	**Expediting Clerk**
COPIES	**3**
DISTRIBUTION	**Contract Manager, Expeditor**
SIZE	**8½ × 11 in.**

PURPOSE

Expediting is defined as the process of ensuring the flow of adequate supplies of material and equipment for continuity of production. The nature of expediting is such that its prime purpose is to serve the functions of information-gathering and communicating. Most expediting systems are proficient at gathering information, but inadequate at communicating.

The nature of the electrical contracting business is such that the Contract Managers *must* be kept informed of all pertinent changes immediately. Therefore, the Expeditor *must* transmit all such information in writing—clearly and completely. The Material Schedule Report was created for this purpose. This memo must be completed in triplicate; two copies go to the Contract Manager, and one copy is stapled to the pink Purchase Order copy and filed by the Expeditor.

The Expeditor must exert all conceivable pressure to force vendors to meet the needed dates. People cannot be tied up on the jobs.

The requirement date for the use of material and equipment is the target that the process of purchasing is destined to achieve, and it is mandatory that it do so. The items that require expediting are material, equipment, tools, and manufacturers' drawings. It would be much more difficult to meet completion dates for most projects, and to manage the job with continuity and dispatch, without considerable expediting—and expediting requires a good set of procedures, because without them the process is extremely frustrating.

MATERIAL SCHEDULE REPORT

Job Name: _____   Date:____ / ____ / ____

Work Order Number: _____   Floor: _____

Purchase Order Number: _____   Class: _____

Contract Manager: _____

Requirement Date: _____

Lead Time: _____

Shipping Date: _____

Delivery Date: _____

Group Material: _____

Shipping Instructions: _____

Delivery Information: _____

_____

_____

_____

_____

_____

_____

_____

_____

Miscellaneous Information: _____

_____

_____

_____

_____

_____

_____

_____

_____

_____

_____

_____

_____

_____

_____

	NAME	**BILL OF MATERIAL (CABLE FIELD MEASURE)**

NAME **BILL OF MATERIAL (CABLE FIELD MEASURE)**

NUMBER **P-3**

ORIGINATOR **Construction Secretary**

COPIES **3**

DISTRIBUTION **Contract Manager, Purchasing Agent, Warehouse**

SIZE **8½ × 11 in.**

PURPOSE When the exact cutting lengths have been determined by the field, they are transmitted to Purchasing through the Contract Manager on the standard Bill of Material for Cable Field Measure. From this Bill of Material the cable will be cut, packaged, delivered, and billed to the job according to standard procedure. Exact cutting lengths of each feeder conductor are determined, by actual measurement, after the feeder conduits are in place.

BILL OF MATERIAL FOR CABLE FIELD MEASURE

Floor Number: ___ Group: *Cable field measure* Lead Time: ___ Sheet Number: *6a*

Job Name: *Empire State Building* Ship to: *Job-site* Date: ___ / ___ / ___

Contr. Number: ___ Engineer: ___ Contract Manager: *V.T.* Work Order Number: *18971*

Quantity Shipped	Quantity Ordered	C L	Item Code	Description		Where Used	Date Required	P.O. Number
	3 pcs.		330423	Wire. Al. 3/0. THW Black	98 ft	LPA	10-31	
				(3—reels)				
	3 pcs.		330423	Wire. Al. 3/0. THW Black	90 ft	PP		39056
	1 pc.		330423	Wire. Al. 3/0 THW Black	87 ft	PP		
				(4—reels)				
	3 pcs.		330425	Wire. Al.250 MCM. THW Black	138 ft	Service		
	1 pc.		330425	Wire. Al.250 MCM. THW Black	125 ft	Service		
				(4—reels)				
	3 pcs.		331408	Wire. Cu.#2. THW. Black	73 ft	Elevator		
	1 pc.		331408	Wire. Cu. #2. THW White	75 ft	Elevator		
				(4—coils)				

The field presents the following information to the Contract Manager on the feeder cutting lengths or calls the information to the Construction Secretary:

✓ *Feeder designation.* This must be the same designation shown on the Feeder Listing.

✓ *Exact length required* (in feet and inches) for each conductor in the feeder.

✓ *Type and size of conductor and type and color of insulation* if for any reason these are different from what is shown on the Feeder Listing; otherwise, the feeder designation is sufficient.

✓ *Packaging*—coils or reels and number of conductors per coil or reel.

✓ *Date required.* This must be at least four working days after the exact cutting lengths are submitted.

BILL OF MATERIAL FOR CABLE FIELD MEASURE

Floor
Number: _____ Group: _____ Lead Time: _____ Sheet Number: _____

Job Name: _____ Ship to: _____ Date: ____ / ____ / ____

Contr. Number: _____ Engineer: _____ Contract Manager: _____ Work Order Number: _____

Quantity Shipped	Quantity Ordered	C L	Item Code	Description		Where Used	Date Required	P.O. Number

NAME	**MATERIAL REQUISITION**
NUMBER	**P-4**
ORIGINATOR	**Contract Manager**
COPIES	**3**
DISTRIBUTION	**Contract Manager, Purchasing Department, Warehouse**
SIZE	**8½ × 11 in.**

PURPOSE   This form is used to order job material that is not on the Bill of Material by item or not in sufficient quantity. It is also used to order expendable material. Most Change Order material requires immediate processing, and is shipped in its entirety from the company warehouse for immediate delivery. This material is listed on the Material Requisition.

Expendable Materials: These are items used in connection with the installation of material and equipment on electrical jobs. They are necessary to the installation work, but for the most part they do not remain as an integral part of the installation work; hence they are expendable. Expendable materials are purchased by the Purchasing Department and are charged to merchandise inventory. Material Orders are written for all these items and are charged to the job work order number. These items are disbursed by the warehouse personnel with the assistance of the toolman, and they do not appear on the Bill of Material.

The following items are included in this category:

Augers, ship, Irwin	Drills, high-speed twist	Paste, soldering
Batteries, dry cell	Drills, masonry	Plaster, patching
Blades, bandsaw	Emery cloth	Points
Blades, hacksaw	Ends, hose	Prestolite, torch sets
Brooms	Files, delta half-round	Punch die sets, Whitney
Brushes, paint	Files, flat	Rags, wiping
Brushes, steel scratch	Files, round	Rasps, file
Burnsmatic	Files, slim taper	Reamers, drill
Cans, oil	Flux, soldering	Rods, steel
Canvas, drop cloth	Gasoline, white	Rope, hemp
Chamois, cleaning	Handles, hammer	Sash cord
Chisels, cold	Handles, pick	Saws, hole
Clamps, hose	Handles, shovel	Soap, pulling
Cloths, wiping	Hose, water	Steel, star bit
Connections, air hose	Jack bits	Steel wool, welding
Couplings, hose	Lines, chalk	Tags, pipe
Cups, drinking	Mandrels, drill	Taps, tape
Dies, bolt	Oakum	Thinner, paint
Dies, conduit threading	Oil, cutting	Tubs
Dope, pipe	Oil, jack	Waste, cloth
Drills, hank, star	Oxygen, tank	Wood, lumber
Drills, high-speed tapered		

MATERIAL REQUISITION

_____ of _____

☐ Will Call
☐ Ship to Job Site
☐ Ship to Warehouse-hold

Job Name: _____

Job #: _____

Written by: _____ Return to: _____ Received by Purchasing: _____
Date

Date Required: _____

Quantity Ordered	Quantity Shipped	Vendor	P.O. Number	Description	Price	Cost Code

NAME	**REQUEST FOR PROOF OF SHIPMENT**
NUMBER	**P-5**
ORIGINATOR	**Expeditor**
COPIES	**1**
DISTRIBUTION	**Vendor**
SIZE	**5½ × 8 in. (two-part) (Copy form at 117% of original.)**
PURPOSE	This two-part card is issued to the vendor to obtain confirmation of the shipping date agreed upon at the time of issue of the Purchase Order. One part is called Request for Proof of Shipment, and the other part, which is detached and returned to the buyer, is called Proof of Shipment. When proof of shipment has been positively established by receipt of the freight bill routing and "pro" number, the Expediter must establish the date of actual receipt and notify the Contract Manager immediately.

REQUEST FOR **PROOF OF SHIPMENT**

JOBBER

JOBBERS P.O. NO.

P.O. NO.

DESCRIPTION

SCHEDULED SHIPPING DATE

PLEASE FORWARD THE ATTACHED POST CARD

WITH COMPLETE INFORMATION ON:

PROOF OF SHIPMENT

This is a vital part of your service to us. Please be prompt.

To Vendor:

FIRST CLASS
PERMIT NO. 0000

FIRST CLASS
PERMIT NO. 0000

VIA AIR MAIL

PROOF OF SHIPMENT

Description:

Jobber P.O. Number:

Customer P.O. Number:

Shipping Date:

Route:

"Pro" Number:

Mfg.:

By:

NAME	**INFORMATION REQUEST CARD**
NUMBER	**P-6**
ORIGINATOR	**Expeditor**
COPIES	**I**
DISTRIBUTION	**Vendor**
SIZE	**5½ × 8 in. (two-part) (Copy form at 117% of original.)**

PURPOSE   With this form, the Expeditor seeks confirmation from the source of supply that the delivery of material will conform with the promised delivery dates and conditions.

In the process of expediting it is essential to adhere to uniform terminology. Some definitions follow:

Requirement date: This is the actual time of need on the job.

Schedule date: The day of need as established on the Activity Schedule.

Activity Schedule: The time bar chart prepared by the processor, showing the sequence of activities according to work schedule.

Lead time: Amount of time in weeks allowed by the Purchasing agent to establish the date of issue.

Expediting date: This is the date that is established by subtracting the lead time from the requirement date. The expediting date determines the time for issuing the Request for Proof of Shipment.

Stock schedule date: The date established for warehouse review of groups of material for each job by subtracting 30 days from the requirement date.

Shipping date: This is the date of shipment from the factory. It is the requirement date less the time required to transport material from the factory to the shop.

DATE: _____

Gentlemen:

Please furnish **immediately** the following information on our

ORDER NO. _____

DATED_____

☐ 1. **Will you ship as required on** _____

_____ ?

☐ 2. Material urgently needed. Give best possible shipping date.

☐ 3. When will you ship balance due?
Give present status of order.

☐ 4. Forward acknowledgment.

☐ 5. Invoice not received.
Send in triplicate.

☐ 6. _____ request for information.
Reply immediately.

☐ 7. Has shipment been made? Give full details.

☐ 8. _____

_____

Please reply on attached card.

Yours very truly,
Purchasing Department

PERFORATION

PURCHASING DEPARTMENT

In reply to your inquiry of

we are pleased to advise as follows:

ORDER NO._____

DATE _____

Date

COMPANY _____

ADDRESS _____

Signed _____

ATTN:EXPEDITOR

POSTAGE WILL BE PAID BY

BUSINESS REPLY MAIL
No Postage Necessary If Mailed in the United States

FIRST CLASS
PERMIT NO.0000

PERFORATION

To:

NAME	**JOB EXPENSE FORM**
NUMBER	**P-7**
ORIGINATOR	**Contract Manager**
COPIES	**2**
DISTRIBUTION	**Contract Manager, Purchasing Department**
SIZE	**8½ × 11 in.**

PURPOSE The services required for a job are ordered on a Job Expense Form, rather than on a Bill of Material. If both services and material are required, the Job Expense Form is still used. This form enables the Contract Manager to order services for a job. Typical Job Expense Form items are:

✓ Heavy hauling.
✓ Crane service.
✓ Trenching.
✓ High-potential testing.
✓ Concrete sawing.
✓ Core drilling.
✓ Telephone service.
✓ Carpentry work.
✓ Painting.
✓ Sound system installation.

If the work is extensive and involved, a subcontract agreement should be drawn up.

The Job Expense Form is prepared in the same way as the Bill of Material except for the Description column. Here the service required is stated in detail, together with the name and address of the company that will perform the work.

Many times the arrangements have already been made by the Contract Manager, and a Purchase Order number has been given. In such cases, "Confirming" is written on the form, and the Purchase Order number is listed in the appropriate column. The Job Expense Form is distributed and filed in exactly the same manner as the Bill of Material.

JOB EXPENSE FORM

Floor Number:_____ Group:_____ Lead Time:_____ Sheet Number:_____

Job Name:_____ Ship to:_____ Date:_____

Contract Number:_____ Engineer:_____ Contract Manager:_____ Work Order Number:_____

Description	Where Used	Date Required	P.O. Number

Requested by:_____

CHAPTER ELEVEN
WAREHOUSE

FORM NO.	FORM NAME	PAGE
W-1	Receiving and Inspection Report	306
W-2	Tool Requisition	308
W-3	Material Disbursement Form	310
W-4	Vendor Return Material Record	312
W-5	Tool Transfer Form	314
W-6	Material Release Order	316
W-7	Shipping Out Form	318

It is the responsibility of the warehouse to maintain a sufficient stock of standard materials, and to provide for initial job shipments of embedded or rough-in material and Change Order material. Most initial shipments require truck trailers for job storage as designated by the Contract Manager, together with job storage facilities and material handling equipment.

The objective of the entire material-handling procedure is to accomplish the following:

✓ Provide the right material to the job at the designated requirement date
✓ Provide proper job facilities for material handling and storage
✓ Make certain that the tools delivered to the field are in good working order, with all parts available

The following processes are involved in the warehousing functions as they relate to material management:

✓ Material storage and disbursement
✓ Shop Delivery Order procedure
✓ Receiving and Inspection Reports
✓ Job facilities for handling and storing material
✓ Tool storage and repair
✓ Job facilities for handling and storing tools

The following equipment is necessary for material handling in both the shop warehouse and on the job:

✓ Plastic bins
✓ A container rack for storing bins
✓ A three-tiered cart on wheels
✓ A heavy-duty four-wheel industrial truck
✓ A two-wheeled trunk carrier-type truck
✓ Metal tote boxes to fit on the three-tiered cart
✓ A tool gang box
✓ A material gang box for cartons

Packaging facilities must be provided to accommodate material-handling needs for the following operations:

✓ Putting up material orders in the warehouse
✓ Transporting material to the warehouse
✓ Receiving and storing material on the job
✓ Carting material to the point of usage on the job
✓ Conveying material back to the shop for subsequent credit

The equipment and containers must meet the following requirements:

✓ Available in uniform sizes and colors
✓ Partially open-ended for visual observation of contents
✓ Stackable
✓ May be fitted to a rack and easily detached
✓ Made of sturdy material
✓ Made of lightweight material

NAME	**RECEIVING AND INSPECTION REPORT**
NUMBER	**W-1**
ORIGINATOR	**Field or warehouse**
COPIES	**4**
DISTRIBUTION	**Warehouse, field, Purchasing Department, vendor**
SIZE	**8½ × 11 in.**

PURPOSE · The Receiving and Inspection Report provides Purchasing with:

1. Evidence of the contents and condition of any shipment to the field so that they may:
   a. Close the expediting file.
   b. Clear the invoice for payment.
   c. Pursue shortages, wrong materials, or damaged items.
2. A means for the field to check for the contents and condition and to record shipments against Bills of Material to afford better material control.

One of several forms constitutes the R&I Report, depending on how the delivery is made:

✓ No R&I Report is required for shipments made to the field from the warehouse, via a company truck, providing the shipment is accurate and in good condition.

If discrepancies occur, make such notes on the packing ticket of the Bill of Material and give it to the Contract Manager immediately.

✓ No R&I Report is required for shipments made directly to the field from vendors' trucks or common carriers, providing a packing ticket is available.

The packing ticket is to be signed and dated by the person receiving the shipment, as well as the freight bill with any discrepancies noted.

If a packaging ticket is not *available*, then fill out an R&I Report, listing the items (by catalog number, type, or description) received, shipper, Purchase Order number, etc. Also note any discrepancies.

RECEIVING AND INSPECTION REPORT

Receiving Tally Number: _____

Item Number	Quantity	Catalog Number	Description of Article
1			
2			
3			
4			
5			
6			
7			
8			
9			
10			
11			
12			
13			
14			
15			
16			
17			
18			
19			
20			
21			
22			
23			
24			

P.O. Number: _____

Vendor: _____ Date Received: _____

Shipper: _____ Shipped from: _____

Via: _____ Waybill Number: _____ Unloaded at: _____

Remarks: _____

Received by: _____ Checked by: _____ Approved by: _____

Matched with P.O. by: _____

Matched with Invoice by: _____ Invoice Number: _____ Date: _____ Amount Approved: _____

NAME	**TOOL REQUISITION**
NUMBER	**W-2**
ORIGINATOR	**Contract Manager, Job Foreman**
COPIES	**2**
DISTRIBUTION	**Warehouse (Tool Room), Job Foreman**
SIZE	**8½ × 11 in.**

PURPOSE

The importance of good tools on an electrical construction job cannot be overemphasized. Tools must be up-to-date—the most modern available—and they must be in good working condition: effective, fast, and safe. Tools must be on the job at or before the time they are needed, and the supply must be plentiful enough so that wireworkers do not have to wait for tools.

Meeting these objectives requires a substantial investment, but it can be minimized with good control. Good control is accomplished by:

✓ Standard nomenclature.

✓ Standard manufacturers.

✓ Accurate tool records.

✓ Effective maintenance and repair.

The term *tools* includes all implements used to accomplish work that are not consumed in the process. That is, tools can be used over and over again to perform the same work.

The Tool Requisition should be prepared in duplicate and filled out completely. Every Tool Requisition should include:

✓ Job name (exactly as it appears on the Active Job List).

✓ Work Order number.

✓ Date ordered.

✓ Date and hour required on the job.

✓ Mark for disburse.

✓ The quantity desired.

✓ Name of the tool. If the name of the desired tool is not preprinted, a description of it should be written in the space provided at the bottom of the requisition exactly as it appears on the Standard Tool List.

✓ Signature of person writing the requisition.

TOOL REQUISITION

Job Name: _____ Work Order Number: _____

Date Ordered: _____ Date Required on Job: _____ A.M. / P.M.

Disburse   Return   Transfer from: _____

Quantity Ordered	Quantity Shipped	Tool Number	Description	Quantity Ordered	Quantity Shipped	Tool Number	Description
___	___	___	Bag, screw, color or type _____				Ladder, step, type _____ height _____
___	___	___	Bag, screw, color or type _____				Ladder, step, type _____ height _____
___	___	___	Bag, screw, color or type _____				Meter, type _____ range _____
___	___	___	Bender, hydraulic type _____ size ____				Meter, type _____ range _____
___	___	___	Bender, manual type _____ size ____				Powder tool, caliber ____ type _____
___	___	___	Bender, mechanical type ____ size ____				Puller, cable, size _____ type _____
___	___	___	Box, gang, (flat) _____				Punch, KO, manual size _____
___	___	___	Box, gang, (upright) _____				Punch, KO, hydraulic size _____
___	___	___	Cart, 4-wheel w/shelves _____				Punch, KO, ratchet size _____
___	___	___	Can, oil, type _____				Rack, contained W/96 stacking boxes ____
___	___	___	Cord, extension wire size and no.__ length ___				Reamer, pipe—size _____
___	___	___	Crimper, wire type _____ size ____				Reamer, pipe—size _____
___	___	___	Cutter, type _____ size ____				Reel, wire, no. of spools _____
___	___	___	Cutter, type _____ size ____				Reel, wire, no. of spools _____
___	___	___	Drill motor ¼ in. _____				Saw, type _____
___	___	___	Drill motor ½ in. _____				Saw, type _____
___	___	___	Drive, power, type _____				Shaft, universal _____
___	___	___	Generator, KW _____ volts ____				Shovel, type _____
___	___	___	Grinder, bench _____				Shovel, type _____
___	___	___	Grips, (kellum) no and size of cond: ____				Spade, type _____
___	___	___	Grips, (Kellum) no and size of cond: ____				Tape, fish type _____ length _____
___	___	___	Hammer-drill small ⅜ in phillips ____				Tape, fish type _____ length _____
___	___	___	Hammer-drill medium (B&D) _____				Tape, fish (vacuum kit) _____
___	___	___	Hammer-drill large (B&D) _____				Threader, 3-way _____
___	___	___	Hammer, electric, type _____				Threader, ratchet, size _____
___	___	___	Hammer, electric, type _____				Threader, ratchet, size _____
___	___	___	Hoist, coffin, weight _____				Vise, portable, size _____
___	___	___	Jack, reel, size of reel _____				Vise, portable, size _____
___	___	___	Jet line tank assembly _____				Wrench, pipe, type _____ size ____
___	___	___	Jet line tank refill _____				Wrench, pipe, type _____ size ____
___	___	___	Ladder, extension length __ ft. _____				Wrench, pipe, type _____ size ____
___	___	___	Ladder, step, type _____ height ____				

Ordered by: _____

NAME	**MATERIAL DISBURSEMENT FORM**
NUMBER	**W-3**
ORIGINATOR	**Warehouseman**
COPIES	**3**
DISTRIBUTION	**Field, Accounting, warehouse file**
SIZE	**8½ × 11 in.**
PURPOSE	On this form is recorded the disbursement of material and electrical equipment for delivery to jobs and over-the-counter sales.

All expendible material must be disbursed using this form. This material is an especially important item because every job uses some expendible material, none of which appears on a Bill of Material.

Expendible material is defined as that which is necessary for the installation work, but for the most part does not remain as an integral part of the structure. The following is an exemplary list of this type of material:

✓ Batteries, dry cell
✓ Blades, hacksaw, bandsaw
✓ Brushes, paint
✓ Cans of oil, paint or water
✓ Cloth, drop, canvas, wiping, emery
✓ Chisels, cold
✓ Clamps, hose
✓ Couplings, hose
✓ Dope, pipe
✓ Flux, soldering
✓ Oxygen tank
✓ Paste, soldering
✓ Rags, wiping
✓ Steel wool, welding
✓ Strings, pulling
✓ Taps, pipe
✓ Wood, lumber

This form has a high degree of versatility and can be used for a number of purposes in the warehouse.

MATERIAL DISBURSEMENT FORM

Number: _____

Material costs shown below are based on our inventory average costs.

Our customers benefit from our ability to purchase in large quantities.

W.O. Number: _____

Date:____ / ____ / ____

Name: _____

Address: _____

Disbursed by:	Cash:	Charge:	Mdse. Returned to Vendor:	Job Credit:	Customer's P.O. Number:	Paid out:

Quantity	Material Number	Description	Sales Price		Cost	
			Unit	Extension	Unit	Extension

Received the above in good order

Total

by: _____

NAME	**VENDOR RETURN MATERIAL RECORD**
NUMBER	**W-4**
ORIGINATOR	**Warehouse worker**
COPIES	**3**
DISTRIBUTION	**Vendor, Accounting Department, warehouse**
SIZE	**8½ × 11 in.**
PURPOSE	This document contains information relative to the return of material or electrical equipment to a vendor or manufacturer.

The form also alerts the Accounting Department to the transaction so that it can:

1. Price out the transaction and make an adjustment to the warehouse inventory account.
2. Price out the transaction and make an adjustment to the job cost record.

VENDOR RETURN MATERIAL RECORD

Number: _____

Vendor: _____ Date: _____

Invoice Number: _____ P.O. Number: _____

Job Number: _____ Foreman/Progress Manager: _____

We are returning the following material to you.

Quantity	Part Number	Description	Unit Price	Total

Reason for Return: _____

_____

_____

Ship Via: _____

Received by: _____

Company Name: _____ Title: _____

Restock Charge: ☐ Yes ☐ No Percentage: _____

NAME	**TOOL TRANSFER FORM**
NUMBER	**W-5**
ORIGINATOR	**Contract Manager**
COPIES	**3**
DISTRIBUTION	**Contract Manager, Job Foreman, warehouse**
SIZE	**8½ × 11 in.**

PURPOSE   On this form, record the transfer of tools from one job to another.

It is impractical for a contractor to maintain enough popularly used tools to provide one for every job and to keep them on hand for the duration of the job.

This form enables the warehouse personnel to maintain:

✓ Current records of the location of tools.

✓ An accurate inventory of all tools on each job at any time.

TOOL TRANSFER FORM

Tool Room use only		
From Location: _____	
To Location: _____	

Job/Crew/Truck Number: _____

Date:_____ / _____ / _____

Number:_____

	Received by (signature): _____	Tool Room Use Only: Disposition
Bar Code Number	**Description**	

Comments: _____

_____

_____

_____

_____

	NAME	**MATERIAL RELEASE ORDER**

NAME **MATERIAL RELEASE ORDER**

NUMBER **W-6**

ORIGINATOR **Contract Manager**

COPIES **3**

DISTRIBUTION **Contract Manager, Purchasing Department, warehouse**

SIZE **8½ × 11 in.**

PURPOSE The Contract Manager uses this form to record instructions to Purchasing to release a group or class of material to a job as it progresses through the various stages of construction.

On sizable jobs requiring large quantities of material, when it is not practical to place all the material on the job at the same time, it becomes advisable to release small quantities at a time in accordance with a predetermined schedule based on the various stages of construction.

Most jobs require only one overall subdivision of work, but large jobs and multistory buildings require more. In addition, important subdivisions of groups represent activities that must be provided for separately.

Each group of material is developed to conform to a stage of construction. Within each group are one or more classes of material. The class of material represents a complete subdivision of work. For most jobs, the group of material represents an activity for use on the Activity Schedule, but a finer breakdown involves classes or items within a class. An *activity* is defined as a stage of construction that must be distinguished from other stages of construction on the Construction Schedule to ascertain that there will be no interruptions in the continuity of the work as a result of the unavailability of an item of material or a piece of equipment. An example of groups of material (allied with stages of construction), and classes of material falling within the groups are the following:

Group Number	Group Name	Class Number	Class Name
06	CABLE	V	FEEDER CABLE
		X	WIRE ACCESSORIES
07	SWITCHGEAR	L	SWITCHBOARD
08	BRANCH CIRCUIT	C	CONDUIT AND FITTINGS
		O	OUTLET BOXES, COVERS, BAR HANGERS
09	WIRE	W	WIRE AND CONNECTORS

The whole system of material classification must be set up in this manner in order to schedule the job as suggested.

RELEASE ORDER

Show this number on all
invoices and shipments.

Date:____ / ____ / ____

To: _____   Ship To: _____

_____   C/O: _____

_____   _____

V/A	Salesperson	F.O.B.	Job Number	Terms	Date Required

Item Number	Quantity	Description	Unit Price	Amount

By: _____

NAME	**SHIPPING OUT FORM**
NUMBER	**W-7**
ORIGINATOR	**Warehouse worker**
COPIES	**3**
DISTRIBUTION	**Shipper, warehouse, Accounting Department**
SIZE	**8½ × 11 in.**
PURPOSE	On this form, record information of shipments made to a manufacturer or to a remote job out of town.

On this form, record information of shipments made to a manufacturer or to a remote job out of town.

This form also alerts the Accounting Department to the need to price out the transaction so as to make adjustments to the following accounts:

1. Warehouse inventory account
2. Job cost record

SHIPPING OUT FORM

All shipments must be ready to ship by 2:30 P.M. Please have a preprinted label made out, and have it with the package.

Job Number: _____   Date:_____ / _____ / _____

Check One: UPS _____   Freight _____

Number of Packages: _____

FOR UPS ONLY

Check One.

Ground: _____

Next Day: _____

2nd Day: _____

Name: _____

Street Address: _____

City: _____ State: _____ Zip:_____

Shipping for: _____

- -

SHIPPING USE ONLY

Date Shipped: _____

Weight: _____

Freight Company: _____

INDEX

Acceleration, Summary of Costs, Form C-3, 261
  procedure, 260
Accident Report, Form M-11, 23
  procedure, 22
Accounts Payable Invoice Master, Form A-2, 49
  procedure, 48
Accounts Payable Invoice Register, Form A-3, 51
  procedure, 50
Accounts Payable Vendor Master (computer screen), Form A-1, 47
  procedure, 46
Accounts Receivable Aging Report, Form A-12, 71
  procedure, 70
Accounts Receivable Job Master (computer screen), Form A-9, 65
  procedure, 64
Activity Schedule Form JP-4, 241
  procedure, 238
Activity Schedule Work Sheet Form JP-3, 237
  procedure, 236

Balance Sheet Form F-1, 87
  procedure, 86
Bid Summary Sheet Form E-21, 169
  procedure, 166
Bill of Material, Form JP-5, 243
  procedure, 242
Bill of Material for Cable Field Measure, Form P-3, 293
  procedure, 292
Branch Circuit Evaluator, Form E-9B, 139
  procedure, 138
Branch Circuit Tabulating Sheet, Form E-16, 153
  procedure, 152
Branch Circuit Takeoff Sheet, Form E-8, 135
  procedure, 134
Branch Circuit Takeoff—Shortcut Method, Form E-9A, 137
  procedure, 136

Budget of Total Sales and Quotas for All Departments, Form M-6, 13
  procedure, 12

Cash Receipts Journal, Form A-13, 73
  procedure, 72
Change Order, Form JM-21, 227
  procedure, 224
Change Order Bid Summary, Form JM-22, 229
  procedure, 228
Check Register, Form, A-4, 53
  procedure, 52
Conductor Tabulating Sheet, Form E-17, 155
  procedure, 154
Confidential Foreman Rating, Form JH-10, 203
  procedure, 200
Contract Manager:
  material control of jobs, 180
  objectives of, 180
  responsibilities, 180
Costs for Engineering, Summary of, Form C-6, 267
  procedure, 266

Delivery Order, Form JM-5, 191
  procedure, 190
Distribution Equipment Takeoff Schedule, Form E-5, 127
  procedure, 126

Embedded Tabulation Sheet, Form E-14, 149
  procedure, 148
Employee Information Record, Form JM-1, 183
  procedure, 182
Engineering Clarification Request, Form C-4, 263
  procedure, 262
Equipment Maintenance Record, MF-4, 113
  procedure, 112

Errata Sheet, Form, E-24, 177
   procedure, 176
Estimate Evaluation Report, Form JM-20, 223
   procedure, 222
Estimate Summary (computerized), Form D-2, 99
   procedure, 98
Estimating, 116
   assumptions, 117
   deviations from standard, 117
   procedure, 117
   takeoff systems, 116
   work schedules for material listing, 116
Estimating Man-Hour Register (computer screen), Form
   M-12, 25
   procedure, 24
Extended and Excessive Overhead, Form C-9, 273
   procedure, 272

Feeder and Busway Takeoff Sheet, Form E-7, 133
   procedure, 130
Feeder Conduit Tabulating Sheet, Form E-15, 151
   procedure, 150
Feeder Cutting Lengths, Form JM-16, 215
   procedure, 214
Field Personnel Time Report, Form JM-2, 185
   procedure, 184
Final Summary of Direct Job Costs, Form C-7, 269
   procedure, 268
Financial Ratios, Optimal Values for Critical, Form F-4, 91
   procedure, 90
Finishing Tabulating Sheet, Form E-18, 157
   procedure, 156
Fixture Schedule, Form JP-7, 247
   procedure, 246
Functional Organization Chart, Form M-1, 3
   procedure, 2

General Ledger Chart of Accounts (computer screen),
   Form A-16, 79
   procedure, 78
General Ledger (computer screen), Form A-15, 77
   procedure, 76
General Ledger Report, Form A-6, 59
   procedure, 58

Home Office Indirect Expense, Allocation of, Form C-8, 271
   procedure, 270

Income and Expense, Summary Form F-3, 89
   procedure, 88
Inefficient Labor Due to Acceleration, Form C-2, 259
   procedure, 258
Information Request Card, Form P-6, 299
   procedure, 298
Invoice, Form A-8, 63
   procedure, 62

Job Closeout Checklist, Form JM-18, 219
   procedure, 218

Job Cost Master (computer screen), Form A-7, 61
   procedure, 60
Job Evaluation Report, Form JM-19, 221
   procedure, 220
Job Expense, Form P-7, 301
   procedure, 300
Job Expense Requisition, Form JP-6, 245
   procedure, 244
Job Factor Evaluation Sheet, Form E-22, 171
   procedure, 170
Job management, 180
Job Processing Check Sheet, Form JP-1, 233
   procedure, 232
Job Registration, Form E-1, 119
   procedure, 118

Lighting Fixture Takeoff Schedule, Form E-4, 125
   procedure, 124

Manpower Chart, Form E-23, 175
   procedure, 172
Material Disbursement, Form W-3, 311
   procedure, 310
Material Inventory, Form M-14, 29
   procedure, 28
Material Release Order, Form W-6, 317
   procedure, 316
Material Requisition and Returned Material Record, Form
   M-15, 31
   procedure, 30
Material Scheduling Change, Form JM-12, 207
   procedure, 206
Memo, Form M-8, 17
   procedure, 16
Memorandum, Form JM-7, 195
   procedure, 194

Outlet Detail and Takeoff Sheet, Form E-10, 141
   procedure, 140
Overhead by Department, Application of, Form M-5, 11
   procedure, 10
Overhead Budget Summary, Form M-7, 15
   procedure, 14
Overhead on Labor and Material, Application of, Form
   M-4, 9
   procedure, 8
Overload Condition Evaluation, Form C-12, 279
   procedure, 278
Overloading, Man-Hours, Lost Due to, Form C-11, 277
   procedure, 276
Overloading, Price Proposal for, Form C-13, 281
   procedure, 280
Overtime Premium Pay, Table of, Form C-1, 257
   procedure, 256

Paid Out Report, Form JM-4, 189
   procedure, 188
Panelboard Schedule, Form JP-8, 249
   procedure, 248

Payables by Job, Form A-5, 57
  procedure, 56
Payables Record by Job, Form A-5, 57
  procedure, 56
Payroll Register, Form A-18, 83
  procedure, 82
Personnel Chart for the Marketing Department, Form
    M-21, 43
  procedure, 42
Personnel Organization Chart, Form M-2, 5
  procedure, 4
Personnel Record, Employee (computer screen), Form
    M-20, 41
  procedure, 40
Power System Takeoff Schedule, Form E-11, 143
  procedure, 142
Pricing and Extensions of the Estimate, Pricing Database
    for (computer screen), Form D-1, 97
  procedure, 96
Pricing Sheet, Form E-19, 161
  procedure, 158
Progress Analysis (computer screen), Form D-4, 103
  procedure, 102
Proposed Material List, Form JP-2, 235
  procedure, 234
Purchase Order, Form P-1, 289
  procedure, 288
Purchasing, 286

Raceway Equipment Takeoff Schedule, Form E-6, 129
  procedure, 128
Recapitulation Summary Sheet, Form E-20, 165
  procedure, 162
Receiving and Inspection Report, Form W-1, 307
  procedure, 306
Reimbursement of Cost of Borrowed Capital, 275
  procedure, 274
Report of Theft, Vandalism, or Damage to Equipment,
    Tools, or Property, Form JM-9, 199
  procedure, 198
Request for Information, Form M-9, 19
  procedure, 18
Request for Proof of Shipment, Form P-5, 297
  procedure, 296
Revision to Bill of Material, Form J-11, 205
  procedure, 204

Safety Meeting Report, Form M-10, 21
  procedure, 20
Sales Budget Calculation, Form M-3, 7
  procedure, 6
Sales Journal, Form A-10, 67
  procedure, 66
Schedule of Fixed Assets, Form A-14, 75
  procedure, 74
Service, Metering, and Grounding Takeoff Sheet, Form
    E-3, 123
  procedure, 122

Shipping Out Form, Form W-7, 319
  procedure, 318
Shop Drawing Log, Form JM-14, 211
  procedure, 210
Shop Drawing Transmittal Form, Form JM-15, 213
  procedure, 212
Special Systems Takeoff Schedule, Form E-13, 147
  procedure, 146
Specification Information Sheet, Form E-2, 121
  procedure, 120
Statement of Profit and Loss, Form F-2, 89
  procedure, 88

Telephone System Takeoff Schedule, Form E-12, 145
  procedure, 144
Termination Slip, Form JM-3, 187
  procedure, 186
Time and Material Billing Register, Form A-11, 69
  procedure, 68
Tool Inventory and Checklist, Form M-16, 33
  procedure, 32
Tool Maintenance and Repair, Form JM-13, 209
  procedure, 208
Tool Master (computer screen), Form MF-2, 109
  procedure, 108
Tool Record, Form MF-3, 111
  procedure, 110
Tool Record Card, Form M-17, 33
  procedure, 32
Tool Register, Form MF-1, 107
  procedure, 106
Tool Requisition, Form W-2, 309
  procedure, 308
Tool Transfer Form, Form W-6, 315
  procedure, 314

Vacation/Sick Leave Request, Form JM-6, 193
  procedure, 192
Vehicle Accident Report, Form JM-8, 197
  procedure, 196
Vendor Return Material Record, Form W-4, 313
  procedure, 312

Wage Evaluation, Employee, Form M-19, 39
  procedure, 38
Weekly Payroll (computer screen), Form A-17, 81
  procedure, 80
Weekly Payroll Report, Form D-3, 101
  procedure, 100
Working Capital Requirements, Form F-5, 93
  procedure, 92
Working Capital, Calculation of Period-to-Period Changes
    in, Form M-13, 27
  procedure, 26
Work Order, Form JM-21, 251
  procedure, 250
Work Stoppage Report, Form C-5, 265
  procedure, 264

About the Authors

Ralph Edgar Johnson is a retired president of Sturgeon Electric Company in Denver, Colorado. He was a pioneer in computerized estimating and job management for electrical contractors. Mr. Johnson is the author of *Electrical Contracting Business Handbook,* and coauthor of McGraw-Hill's *Successful Business Operations for Electrical Contractors.*

Gene Whitson is a former vice president of Cannon-Wendt Electric Company in Phoenix, Arizona. He specializes in industry statistics and business controls, and was a leader in the computerization of the electrical contracting business. Mr. Whitson is coauthor of McGraw-Hill's *Successful Business Operations for Electrical Contractors.*